POLITICAL ECOLOGY AND THE ROLE OF WATER

For the Sa'dah people,
with a great many thanks for their hospitality and generosity

Political Ecology and the Role of Water

Environment, Society and Economy in Northern Yemen

GERHARD LICHTENTHÄLER
School of African and Oriental Studies
University of London, UK

Routledge
Taylor & Francis Group

LONDON AND NEW YORK

First published 2003 by Ashgate Publishing

Published 2016 by Routledge
2 Park Square, Milton Park, Abingdon, Oxfordshire OX14 4RN
711 Third Avenue, New York, NY 10017, USA

First issued in paperback 2016

Routledge is an imprint of the Taylor & Francis Group, an informa business

British Library Cataloguing in Publication Data
Lichtenthäler, Gerhard
 Political ecology and the role of water : environment
 society and economy in Northern Yemen. - (King's SOAS
 studies in development geography)
 1. Water-supply - Yemen - Sa'dah Basin 2. Political ecology -
 Yemen - Sa'dah Basin 3. Sa'dah Basin (Yemen) - Social
 conditions
 I. Title II. King's College, London III. University of London.
 School of Oriental and African Studies
 333.9'1'0095332

Library of Congress Cataloging-in-Publication Data
Lichtenthäler, Gerhard.
 Political ecology and the role of water : environment society and economy in northern
 Yemen / Gerhard Lichtenthäler.
 p. cm. -- (King's SOAS studies in development geography)
 Includes bibliographical references and index.
 ISBN 0-7546-0908-1 (alk. paper)
 1. Water-supply, Rural--Yemen--Sa'dah (Province) 2. Political ecology--Yemen--Sa'dah
 (Province) I. Title. II. Series.

TD313.Y45 L53 2002
363.6'1'095332--dc21

2002028102

ISBN 13: 978-1-138-27738-0 (pbk)
ISBN 13: 978-0-7546-0908-7 (hbk)

Contents

List of Figures

List of Tables

Preface

I had waited for some time for a lift into the tribal areas of the Sa'dah region. Finally, a loaded Toyota pick-up pulled up. Behind the steering wheel on a high pillow sat Musa, looking like a ten-year-old who had skipped school. Musa was only four and a half feet tall but thought he was about 15. We fixed a few national flags on the Toyota and started off.

At the first checkpoint my small friend did an excellent job of convincing two army men why his car did not need number plates or any papers for the items he was taking to the mountains. At the second check-point the soldier who stopped us looked Musa's age but was even shorter, having difficulties even looking through the car window. His authority was definitely enhanced by an automatic rifle.

Musa had a precise idea of how the national flags on our pick-up should blow in the wind. Each time one of the banners got into disorder he stopped the vehicle and usually asked me to restore its image to one of honour and glory. He told me that he had been driving a Toyota up and down the mountains for four years. He must have been 11 years old when he started his career, and hardly able to reach the pedals. Once he had raced to his village in record time with 20 goats on the back and three passengers in front. Now, at 15, he had higher ambitions. He wanted to become a learned man – at least a doctor – as he considered his brothers to be since they owned a pharmacy.

After two hours Musa bought a tin of tomato paste, a tin of fish and some bread. We put the opened tins between us on the front seat, and said 'Bismillah' – 'In the name of God' – and ate. Soon after we attempted the treacherous ascent to the mountainous plateau of Barat, home of the Bakil tribes of Dhu Muhammad and Dhu Husayn. There the road was blocked for several days by a drilling rig, which had rolled over on the narrow mountain track. The owner had succeeded in smuggling this heavy piece of equipment across the border passes, only to lose it on the track up to Barat. He took his fate with an ease and patience that I have only seen amongst Muslims. Allah had willed it. – Who was he to question the Omniscient? 'Al Hamdulillah' – 'Praise be to Allah' – was all he said. A few days later, and with the help of many people, his drilling rig was saved and moved off the mountain track. This happened during the 'fat' years, in 1984, when I first visited the Sa'dah basin.

Almost a decade later, in 1996, I met one of Sa'dah's influential actors at the site of his latest tube well. The sheer noise of the huge compressors powering the rotary drill transformed our initial conversation into a shouting exercise. By the time we moved back to his house for lunch, drilling had reached about 110 meters depth with no indication that the groundwater level was imminent. The 'lean' years had begun.

Sa'dah and its Environs

The Sa'dah basin is the main focus of this book. However, the area is no isolated social, economic or political entity but relates to the surrounding tribal areas in many different ways. Developments in the politicised environment of the Sa'dah basin are better understood by looking at the environmental, social, economic and political characteristics, which define the region as a whole.

I first visited Juma'ah, Munabbih and Razih (of the Khawlan b. Amir tribes) in 1984. In 1986, I made a longer excursion to the east of Sa'dah. It included areas of Wadi Amlah of Al Salim (Wa'ilah, Hamdan-Bakil), a wadi through which most of the smuggled goods were trucked into the Sa'dah basin.

During 1989, while working and living in the Sa'dah basin, I witnessed tribal rivalries and competition for political and economic power. The shoot-out, which happened inside the grounds of the government's main offices, left a number of men dead and many others seriously wounded. This and many similar incidents exposed the limits of central government in the area at the time.

First hand experience of the 1994 Yemeni war and the escalation of tensions between Yemen and Eritrea provided valuable perceptions about food self-sufficiency.

In 1995 a journey to the east of Sa'dah included areas near Najran, at the Saudi border. Contact with a group of Bedouins informed my perceptions of the tribal-political economy at the fringes of the so-called Empty Quarter. Approaching the oasis of Ma'rib from the east, many days later, significant information was volunteered about the politics and expansion of groundwater irrigated agriculture in that area.

During fieldwork for this book I visited a number of areas around the Sa'dah basin. Two trips to Kitaf (Wa'ilah, Hamdan-Bakil), in February and October of 1996, yielded interesting perceptions about the trade in livestock. An incident during the last visit facilitated contact with some of the area's tribal-political actors. Since the late 1970s population pressure as well as land and/or water scarcity has resulted in people from areas near to Sa'dah migrating to the basin. I felt it important to visit these areas in order to understand the social, economic and political push factors.

The Bakil area of Al Ammar, about 35 km south of Sa'dah was explored in October of both 1996 and 1997. A cordial relationship with the tribe's shaykh made it possible to talk openly about many politically sensitive issues and developments relevant to the study.

Al-Mahadhir (Khawlan b. Amir), a small basin just south of Sa'dah, was visited in the autumn of 1996. Since its completion in 1979 the road from San'a to Sa'dah runs through al-Mahadhir. The trip revealed the adaptive capacity of its population. Significant clues were first gained here about the relationship between rights to land and rights to runoff.

Twice, in 1996 and 1997, I went to Bani Uwayr, towering just to the west of al-Mahadhir. This barren and resource scarce plateau has been isolated not only by its geographical characteristics but by drawn-out tribal feuds. Nevertheless, or perhaps because of these factors, Bani Uwayr has produced some of the most successful businessmen and entrepreneurs, all of whom now live in the Sa'dah basin.

Culture and Context

Fieldwork in the Sa'dah basin covered a period of four years, between 1996–1999. Contacts and friendships ensured that research between 1996 and 1999 could be pursued in spite of the prevailing security conditions during that time. Over the past decade kidnapping and car hijacking has become a problem associated, in particular, with the tribal areas of northern Yemen. As a result foreign visitors have periodically been restricted from travelling to the Sa'dah region. During the summer of 1998, for instance, foreign workers travelling from Sa'dah to San'a counted no fewer than 13 tribal and government roadblocks as well as three attempts to have their vehicle hijacked.

On arrival at the capital's airport on January 17th I learned that a number of friends and former colleagues had just been kidnapped while driving from Sa'dah to San'a. In spite of the army closing the 245 km long stretch to foreigners as a result, a number of cultural and linguistic factors as well as other favourable circumstances ensured that I reached Sa'dah the following day. The kidnapping incident even added qualitative data to the study. In the company of the resident director of the NGO, for whom the kidnapped worked, I was able to meet many tribal leaders over the next 16 days. As an aside to the mediation efforts conversations after lunch naturally shifted to topics and concerns related to water issues. I am very thankful for the enormous hospitality, generosity and openness of the Sa'dah people. Throughout my years among the tribes of the region I have experienced freedom of movement and a sense of security and protection.

Fieldwork benefited enormously from the long association with a medical team based at Sa'dah's Republican Hospital, which, over the past 25 years, has won the respect and support of individuals and communities from all the different tribal groups. This fact facilitated access to farms, individuals and groups from all the areas in the Sa'dah basin. Farms were visited during the morning, when most of the work is done. The many occasions when breakfast and/or lunch were taken with farmers and their families proved not only nutritious but extremely informative for understanding the high values attached to locally produced food. An appreciation of the area's cuisine also helped to understand farmer's crop choices.

Most adult males of whatever occupation chew qat during the afternoon hours in the company of friends, relatives, neighbours, or with those they need to meet for a specific purpose. The afternoon qat chew provided an extremely beneficial social setting in which to discuss issues and questions that had come up as a result of participant observation in the morning. The number of men present during qat sessions varied from less than 10 to more than 30. Usually, participants were from the same village or area. But on many occasions they also included men from other tribes and from areas further away. Therefore, much of the information obtained during qat sessions represent a non-scientific sample of perceptions and responses from the Sa'dah area. On many occasions men with detailed knowledge about cross-border trade and smuggling were present and freely shared their experiences. During other times I found myself sitting next to leading tribal personalities from neighbouring regions. Discussions with them helped to situate the study in its wider socio-political and socio-economic context. Moreover, questions frequently triggered debate among the men and subsequently uncovered important side issues. And finally, during the three-hour qat chew there were also long spells when the interests of the foreigner were no longer the focus. This allowed for a quick analysis and the formulation of new sets of questions. Importantly, participation as a fellow qat chewer and silent listener helped to gain a broader understanding about the daily concerns of Sa'dah's tribal communities. Reflecting Kopp's experience in the late 1970s (Kopp 1981:27) best results were achieved generally by participating in qat sessions of small and socially homogeneous groups where people felt free to voice their cóncerns, perceptions and experiences.

Gerhard Lichtenthäler
Neunkhausen
Germany
2002

Author in discussion with tribal men on Bani Uwayr

Acknowledgements

I would like to acknowledge the Sa'dah people who, without exception, received me with great generosity and hospitality and who generally felt at liberty to disclose culturally sensitive information. I am greatly indebted to Drs Truus and Huib Wierda (Sa'dah), who for the past 25 years have sacrificially served the people of the Sa'dah region. The study could not have been done without their friendship and support. I am also grateful to Dr and Mrs Ian Levett (Sa'dah) and to Gerard and Anneke Verbeek for their sustained interest and help. Christopher Ward, Tony Turton, Shelagh Weir, Jac Van der Gun, and Drs Ghulam Farouq (London), Chris Handley (Taiz), Eric Partick and Stefan Kohler are gratefully acknowledged for stimulating discussions, constructive critique and for introducing me to appropriate concepts. I am grateful to Dr Paul Dresch and Professor Tim Unwin for critically reviewing the book. John Latham (FAO) provided the 1998 satellite image, while Yasir Mohieldeen generously assisted with GIS and image processing. I would like to thank Volker Manz for encouraging me to study at SOAS, Professor Keith McLachlan for arousing my interest in the geography of the Middle East, Dr Kate Zebiri for her advice and support during my time at SOAS, and Dr David Zeidan for his friendship and example. The people of Trinity Lutheran Church and also Dr John and Mrs Dorothy Davies encouraged and supported me and my family throughout my time at SOAS, while Commit Trust generously paid the fees for the first year. In addition, SOAS Scholarship Committee and the University of London Research Fund both provided money to conduct fieldwork. I am grateful to the staff of the Geography Department, SOAS, for helping me to navigate through academia. Most of all I would like to thank Debra, my wife, for her unfailing love and support and my supervisor Professor Tony Allan for his productive and efficient supervision and for his outstanding support and encouragement.

List of Abbreviations

AREA	Agricultural Research and Extension Authority
b.	bin (Arabic: son of) e.g. Dirdah b. Ja'far)
CWR	Crop Water Requirements
EC	European Commission
GDH	General Department of Hydrology
HWC	High Water Council (1982–1995)
IDRC	International Development Research Centre
IIMI	International Irrigation Management Institute
MAWR	Ministry of Agriculture and Water Resources
MEW	Ministry of Electricity and Water
MOMR	Ministry of Oil and Mineral Resources
NGO	Non-Governmental Organisation
NWRA	National Water Resources Authority (since 1995)
O&M	Operation and Maintenance
SOAS	School of Oriental and African Studies
TNO	Institute of Applied Geoscience, Delft, The Netherlands
TOR	Terms of Reference
UNDP	United Nations Development Programme
WDM	Water Demand Management
WRAY	Water Resources Assessment of Yemen
WS	Worldwide Services (NGO)
YR	Yemeni Riyal (Currency)

1. The transliteration employed for the purpose of this book is a simplified version of Hans Weir's Arabic – English dictionary. The Arabic *hamzah* glottal stop as well as the *ayn* are represented by the symbol ' but only where they appear in the middle of a word.
2. Between 1996 and 1999, the period of this study, the exchange rate for the Yemeni Riyal varied from 120–140 YR = $1US. 125 YR has been used.

Definitions

For the purposes of this book, the following definitions will apply:

Aquifer. An underground stratum capable of storing water and transmitting it to wells, springs, or surface water bodies.

Depletion defines the withdrawal of water from surface or groundwater bodies at a rate greater than the rate of replenishment (recharge rate).

Groundwater mining describes a condition when withdrawals are made from an aquifer at rates in excess of net recharge.

Adaptive capacity is the sum of social resources that are available within a society that can be mustered in order to counter effectively an increasing natural resource scarcity (Ohlsson, 1998 1999). There are at least two distinct components to adaptive capacity (Turton 1999: 25). The structural component comprises the sum of institutional capacity (including financial capacity) and intellectual capital, which allows for the generation of alternative solutions such as water demand management strategies introduced by technocratic elites. The social component is defined as the willingness and ability of the social entity to accept these technocratic solutions (such as water demand management strategies) as being both reasonable and legitimate.

Allocative efficiency is the second method available to increase the returns to water, which is based on the notion of a rational choice as to which activity would bring the highest return to water (Allan 1998: 3). Allocative efficiency is economically rational but can be politically stressful, so it tends to be avoided by politicians and can be understood as being the 'more jobs per drop' option (Allan 1999*b*:7) two distinct types of allocative efficiency in existence that need to be clarified however, as each represent a different level of political risk. *Inter-sectoral allocative efficiency* can be defined as allocating water away from one economic sector (usually agriculture) because of an inherent low return to water, to another (usually industry) because of an inherently higher return to water. This is politically very risky (Turton 1999:17) as was discovered by President Jimmy Carter when he tried to introduce a new policy called 'realistic water pricing' in the USA (Reisner 1993: 323). *Intra-sectoral allocative efficiency* can be defined as allocating water within a given sector, usually at the level of a production unit (farm or factory), away from production that has a low return to water to production with a higher return to water. This is usually politically less risky [see also allocative efficiency].

Ecological marginalization is one of the end-products of resource capture, whereby those to whom access is denied become marginalized socially, politically and economically (Homer-Dixon and Percival 1996:7). This becomes significant within the context of a developing state where either the economy or the government usually lacks the capacity to provide for those people who are marginalized.

A **first-order scarcity** is a scarcity of a natural resource such as water or land (Ohlsson 1998; 1999) [see also second-order scarcity].

Natural resource reconstruction exists when a social entity can effectively introduce water demand management, specifically by re-allocating water from one economic sector to another (Allan and Karshenas 1996: 127–8). Thus, natural resource reconstruction needs principles of allocative efficiency to be applied in order to become a reality. For this to take place effectively, the second (social) component of adaptive capacity must be present (Turton 1999: 29) and functioning.

Productive efficiency is one of the two methods available to increase the return to water, usually involving improvements to the efficiency of water delivery in irrigation systems (Allan 1998: 3) or to other consumers. The important aspect in this regard is that it does not involve a change in the overall water-use paradigm by allocating water to alternative economic sectors. This makes it a politically favoured but sometimes ineffective option as it fails to take advantage of the gearing ratios with respect to return to water that alternative economic sectors offer. This can be thought of as the 'more crop per drop' option (IIMI 1996) and is sometimes referred to as 'end user efficiency' (Turton 1999: 22).

Resource capture is the process by which powerful social groups shift resource distribution in their favour (Homer-Dixon and Percival 1996:6) over time. This is particularly relevant under conditions of extreme water scarcity where access to a critical natural resource like water gives considerable advantage to those who control the access and allocation of that resource. This serves to politicise water further by decreasing the level of legitimacy, introducing elements of mistrust which undermine the water demand management strategies being proposed by technocratic elites (Turton 1999: 13).

Return to water refers to the value of the product being produced from a given quantity of water. This implies that there are at least three fundamental aspects concerning the notion of 'value', which need to be recognised. Firstly, the concept can be understood in terms of an economically rational approach, naturally favouring the production of goods with the highest economic yield. This may be economically rational, but socially or politically irrational, and consequently not implemented. Secondly, the concept can be understood in terms of a culturally rational approach, which would favour water allocation to an economically nonviable but culturally essential activity. Water as a status symbol can be seen in this regard. Thirdly, the concept can be understood in terms of a politically rational approach, which may be economically irrational but politically necessary and feasible. The concept of return to water is thus highly complex and effectively defies a simple definition, which is probably why other authors seem to have avoided the task.

Sustainability has three dimensions – social, economic and environmental. Society, the economy and the environment must be able to articulate their interests in the policies that determine resource allocation. A sustainable environment can be achieved if social and economic sustainability have also been achieved. Sustainable development can be operationally defined as taking a long-term view on water-use patterns; not compromising the future in pursuit of the present; recognising the need to involve people; and emphasising the quality of life and living systems in the definition of development (Morris 1996: 230). Sustainable development therefore involves a switch in emphasis from supply-management (which attempts to meet rising demands by abstracting more water from a depleted resource base) to demand management (which attempts to reduce consumption by increasing efficiencies and developing alternatives) over time.

A *second-order scarcity* is a scarcity of adaptive capacity (Ohlsson 1998; 1999). This is manifest as the inability of a society to muster sufficient social resources to counter effectively the increasing natural resource scarcity. The operative word here is 'effectively'.

Water demand management (WDM) is a policy for the water sector that stresses making better use of existing supplies, rather than developing new ones (Winpenny 1997: 297). WDM can be managed in many different ways (Westerlund 1996: 155) so there is no given strategy that is universally applicable. This means that local factors such as culture need to be considered by technocratic elites when developing solutions. In general, WDM implies measures and practices including education and awareness programs, metering, water pricing, quantitative restrictions, and other devices, to manage and control the demand for water.

1 Introduction

Population growth, poverty and problems associated with common property resource management have been common themes when trying to explain overexploitation and degradation of natural resources of the countries of the South (Hardin, G. 1968). However, insufficient attention has been paid to how traditional political relations and local perceptions affect natural resource capture and resource allocation. This is especially evident with respect to groups and communities at the political and geographical peripheries of state influence and control for whom self-identity is constructed around notions of autonomy and food self-sufficiency.

Purpose and Scope of the Book

The following chapters address water resource allocation and management, the main aim being to investigate the socio-economic and political contexts which influence approaches to and determine practices of water management, taking the particular example of the tribal communities in the Sa'dah basin of northern Yemen. It seeks to analyse and to understand the politics of environmental change, with particular reference to groundwater resource degradation, within the conceptual framework of 'political ecology'.

Political Ecology

Bryant (1991: 165) defines political ecology as an '[i]nquiry into the political sources, conditions, and ramifications of environmental change'. He puts forward the view that political ecology 'reflects above all a reasoned argument that to understand, as well as to be in a position to solve, the Third World's environmental problems is to appreciate the ways in which the status quo is an outcome of political interests and struggles. Indeed, it is to acknowledge the existence of a politicised environment in the Third World in which power relations play a central role. Thus politics and environment are everywhere interconnected' (Bryant 1997: 9). Consequently, Bryant maintains that political ecology 'focuses on the interplay of diverse socio-political forces, and the relationship of those forces to environmental change' (Bryant 1991: 165).

Political ecology emerged as a distinct research field only during the early 1980s and 'in response to the perceived apolitical nature of the mainstream literature' on 'sustainable development'. Political ecologists argue that most of these studies de-politicise ecology by ignoring the political dimensions of environmental change (Bryant 1991:5; 1997:164).

Political ecology integrates the 'concerns of ecology and a broadly defined political economy' (Blaikie and Brookfield 1987: 17 as cited in Bryant and Bailey 1997: 20) in an attempt to investigate human-environmental interaction. Approaches to the analysis of environmental issues vary and have been summarised as follows (Bryant and Bailey 1997: 21–24):

- Problem specific; analysis of a specific environmental problem to understand human impact on the environment.
- Concept specific; focus on a concept such as 'sustainable development'.
- Regional focus, approaching a political-ecological problem within the context of a specific region.
- Discourse centred; examining issues of political ecology as they relate to socio-economic characteristics such as class, ethnicity and gender.
- Actor-oriented; stressing the importance of understanding political-ecological conflicts as well as co-operation 'as an outcome of the interaction of different actors pursuing often quite distinctive aims and interests' (Bryant and Bailey 1997: 23f).

These different approaches need not be mutually exclusive, indeed they might be combined in research by political ecologists. However, for the purpose of this study the actor-oriented approach will be adopted to investigate the changing dynamics of 'interests' and 'power' of selected actors within the context of groundwater over-abstraction in the Sa'dah basin.

In an attempt to explore the politics of environmental change within the framework of an actor-based approach ecologists ask three main questions (Bryant 1997: 11; Bryant and Bailey 1997: 39):

- 'What are the various ways and forms in which one actor seeks to exert control over the environment of other actors?'
- 'How do power relations manifest themselves in terms of the physical environment?'
- 'Why are weaker actors able to resist their more powerful counterparts?'

The three questions above relate directly to the three main areas of inquiry of the Sa'dah study which seeks to:

- investigate the extent to which environmental degradation, and especially the unsustainable mining of the Sa'dah basin's groundwater resources from the mid-1970s until the mid-1980s, can be explained as the outcome of unequal power relations, political interests, and the changing ability of actors to control or resist other actors.

- explain how groundwater allocation and management in Sa'dah's politicised environment are driven by tribal-political notions of power, knowledge and interests and by perceived and/or real socio-political and socio-economic values. Both, over-abstraction and conservation of Sa'dah's groundwater resources, can be understood in terms of the ability of actors to control and/or resist other actors.
- analyse the extent to which the social and economic 'costs' resulting from groundwater depletion are shared and born unequally by different actors in the study area, which is one of the main assumptions made by political ecologists (Bryant and Bailey 1997).
- explore the comparative social, economic and political resourcefulness of selected actors and communities to adjust to water scarcity while evaluating the capacity of the tribal-political system to address issues of 'sustainability' and equity in the context of groundwater use.
- indicate the socio-political and socio-economic potential and constraints at work in politicised environments such as the Sa'dah basin, to apply measures and principles which are perceived as economically sound and ecologically sustainable by analysts and policy makers informing and guiding national governments.
- demonstrate that in the politicised environment of the Sa'dah basin, where actors display a remarkable ability to resist control, it has not been state regulation but community participation, indigenous solutions and local initiatives which have determined the degree to which measures to improve water use efficiency can be achieved.

The Sa'dah Basin – a Politicised Environment

Central to the idea of a politicised environment is the recognition that environmental problems cannot be understood in isolation from the political and economic contexts within which they are created (Bryant and Bailey 1997:28).

power is at the heart of this politicised environment (Bryant and Bailey 1997: 47).

Politically, economically and in terms of its relation and proximity to Saudi Arabia the Sa'dah basin merits investigation since it reflects a highly politicised environment shaped by power relations and interests, expressed as control and resistance.

Power and Control

Tribal-customary laws pertaining to the allocation and management of the Sa'dah basin's scarce surface water resources will be defined for the period up to the early 1970s, revealing the relative power of one actor over the environment of another. The power of down-stream users with established rights to runoff water effectively controlled and limited any change or development – agricultural or other – of the runoff area, which, in most cases, belonged to other actors upstream. Resulting tensions were exacerbated by the fact that the Sa'dah basin is shared between communities identifying with a number of tribal groups, including the two main rival confederations of Hashid and Bakil. Groundwater exploration and privately owned tube-wells offered one possible way to resist control by other actors. Furthermore, privately owned wells also meant an end to drawn-out and costly tribal conflicts over scarce surface water resources. Moreover, groundwater development promised secure water supplies leading to greater autonomy and food self-sufficiency, ideals which are highly valued in Sa'dah's tribal society. However, the extensification of groundwater irrigated agriculture was, with a few exceptions, impossible since a) most land was communally owned and managed and b) extensification was blocked by the power and control exercised by down-stream owners of runoff rights. A settlement negotiated by a religious scholar in 1972 became a mile-stone decision which opened the way for the expansion of agricultural development and consequently led to the mining of the Sa'dah aquifer. It stipulated that a community must give up half of the area of its grazing land to those owning rights to the run off from it. However, if the owners of run off preferred the run off to receiving half of the land, no development could take place. The scholar's arbitration was accepted unanimously by all the tribes. It consequently altered the control and balance of power with regard to land and water resources in the Sa'dah basin. But, what is significant in the context of this study, the solution was driven, to a large extent, by the emergence of another, potentially powerful, actor – the state. Privatisation of communal lands was this 'solution' and the consequent extensification of groundwater-irrigated agriculture can be understood as direct/indirect results of this change in Sa'dah's politicised environment.

Autonomy and Resistance

Sa'dah is the historical, religious and political centre of Zaydi Islam. During the revolution in 1962 that ousted the Imam of Yemen and the consequent years of civil war parts of the Sa'dah area remained a last bastion and stronghold of Royalist values and opposed to the creation of the Republic. Political autonomy and resistance to state control have remained core values and objectives of many actors and tribal communities. Attempts by the state to establish control over the Sa'dah region have had direct repercussions on groundwater development. As the government tried to establish control over the province, tribal groups feared that the state was out to claim their common land under the pretext of military training, national security and 'development'. In response, communities mobilised quickly and privatised their communally owned areas in an attempt to resist the power and interests of the state. Those with holdings too large for groundwater-irrigated agriculture sold land to members of other tribes – something deemed shameful according to tribal customs. Land sales, in turn, triggered a wave of migration into the basin. For economic and political reasons it became very lucrative to live near the booming tribal markets of the Sa'dah basin. Some, indeed, made fortunes while the hopes of others to diversify their livelihoods away from agriculture were dashed by subsequent political developments and power struggles. Notions of power and interests, control and resistance explain expansion and development of groundwater-irrigated agriculture.

Trade and Transfer

The region's proximity to Saudi Arabia and its social and economic links and networks, which extend beyond their disputed political boundary, explain economic activities. During the oil-boom years of the late 1970s and early 1980s actors in the Sa'dah basin benefited disproportionately, compared with the rest of Yemen, from the unprecedented flow of goods and services across the Yemeni-Saudi border. Attempts by the state to control the highly lucrative cross-border trade during the early 1980s forced many back into agriculture. However, cross-border trade has remained a preferred livelihood of many and is symbolic of the ability of grass-roots actors to resist the power and control of the state.

Values and Vulnerabilities

Crop choice and water use patterns are not driven by purely economic considerations alone: rather, they are determined by social, economic and political values. Political and economic instability, caused and exacerbated by the Gulf crisis, the 1994 civil war, the recent dispute with Eritrea and the Saudi-Yemeni border issue are factors, which have fostered a climate of fear and uncertainty. Moreover, the majority of actors in the study area consider the state as corrupt and lacking in credibility. These perceptions help to perpetuate tribal notions of autonomy and food self-sufficiency. Ultimately, crop choice and water use patterns signify perceptions of power and resistance.

Resources and Resourcefulness

Power relations determine the need, ability and scope of various actors and communities to adjust to water scarcity. Indications are that the costs of groundwater mining are borne unequally by different actors. In the process, actors with relative social, economic and political power, and those with access to 'knowledge' use their position to gain control over the environments of others. At the same time the resourcefulness of grass-roots actors to resist the control of powerful actors and to adjust to scarcity is considerable. In both cases, perceptions of control and resistance drive unsustainable forms of groundwater management and suggest, so far, a 'tragedy of the commons' scenario. However, as has been shown in a variety of different contexts (Ostrom, 1990), communities are able to manage their natural resources in sustainable ways without the control of the state. In Sa'dah's politicised environment and among its tribal communities, which have a long tradition of managing their own water, resources one thing seems certain. It is not state control but community participation, indigenous solutions and local initiatives that have determined the degree to which sustainable management of the basin's vital groundwater resources can be achieved.

2 Sa'dah's Politicised Environment

Introduction

While political ecology provides a useful overall analytical framework for the Sa'dah study a number of other approaches and concepts will be employed to analyse the changing dynamics of water management in Sa'dah's politicised environment.

Allan (1999a:2) has long argued for an interdisciplinary approach to water issues. Within the context of Yemen the need for holistic research has also been demonstrated by Handley (1999:285). While many consider interdisciplinary scientific research 'essential, basic, inescapable, comprehensive, properly holistic and powerful' it has also been noted that such an approach can be conceptually demanding and operationally difficult (Allan, SOAS Geography Dept Research Seminar 1996–97). Nevertheless, the interdisciplinary component of geography, in particular, is considered strong because of its focus on 'relationships rather than on the profile and nature of discipline defined phenomena and processes'. In addition, Allan emphasises that an effective interdisciplinary study 'reflects the interdisciplinary nature of the real world' (ibid).

In the following study a few selected social, political and economic concepts will be incorporated in the Sa'dah study in order to provide a deeper understanding about why individuals and communities manage and allocate their water resources in the way they do.

Adaptive Capacity

The emerging concept of adaptive capacity (Turton and Ohlsson 1999) is useful for understanding how Sa'dah's politicised environment affects, empowers and/or constrains the adaptive capacity of its tribal communities.

It has been shown that societies cope in different ways with water scarcity (Allan and Karshenas 1996). The authors suggest that during the initial phase of development economies tend to deplete their stock of environmental resources. However, this process of resource degradation need not result in a Malthusian catastrophe or a situation where a particular environmental resource base is irreversibly damaged. Depending on whether a society is endowed with sufficient adaptive capacity resource degradation can be slowed down and even reversed resulting in what has been termed natural resource reconstruction (Allan and Karhenas 1996:127f).

Recent studies by Turton (1999:1) and Turton and Ohlsson (1999:2) argue that the adaptive capacity of a society should be considered as a resource. In this context Ohlsson makes a clear distinction between what he terms a first-order scarcity of natural resources and a second-order scarcity of social resources. It follows, therefore, that resource scarcity is not only defined by a lack of environmental capital but by the ability of a social and/or political entity to adjust and cope with this condition. Following this argument a further differentiation has been made between water scarcity and water poverty. While water scarcity pertains to the natural endowment of a society's environmental stock water poverty refers to the inability of a society to adjust to this condition. In other words societies facing water scarcity do not face the predicament of becoming water poor provided that adequate levels of adaptive capacity allow them to adjust and restructure their economies appropriately.

Figure 2.1 visualises the theoretical model first developed by Turton. It indicates the various components and links that determine the level of adaptive capacity.

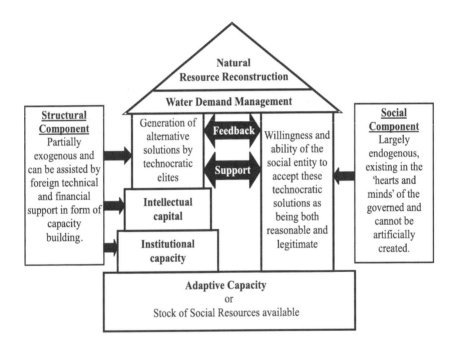

Figure 2.1   Adaptive capacity and natural resource reconstruction

The model resembles a house in which adaptive capacity forms the all-important foundation (bottom block). The top of the roof (triangle) is termed natural resource reconstruction and indicates the goal and the final phase of the process, namely, the sustainable use of water resources. This goal is achieved through water demand management (WDM). However, success of WDM, according to the model, is determined by two supporting pillars. The left pillar, labelled 'structural component' indicates support, expertise and advice that is available through mainly external players such as NGO's, technocrats, and international agencies such as the World Bank and the UNDP. Here, 'institutional capacity', is seen as the basis of the structural component. Effective institutional arrangement will accommodate and support the next critical building block, 'intellectual capital'. Given the complexity of interests at work developing viable strategies and policy options it will be necessary to weave multidisciplinarity into a society's social resources. The top block of the left pillar indicates the expected output, the generation of alternative solutions by technocratic elites, whether locals or foreigners.

The pillar on the right side of the model represents the 'social component' of adaptive capacity. It is increasingly understood that a successful implementation of WDM strategies resulting in natural resource reconstruction depends to a large extend on two main issues of the model's social component. First, stakeholders who are expected to implement measures of WDM must show willingness and ability to accept the proposed measures. Turton points out (1999:27) that these preconditions are unlikely to be met where governments are perceived as corrupt and/or lacking legitimacy. Similarly, perceptions of resource capture by the state or by actors seen as linked to the regime are unlikely to foster willingness and ability among target groups. A second issue is also critical and relates to whether the envisaged WDM strategies are culturally appropriate and socially acceptable.

The arrows between the structural and social component indicate the importance of constant feedback and support. Solutions and strategies, especially when developed with the help of foreign support, need to be adapted to local situations. Community participation and community consultation is crucial for fostering legitimacy and trust.

The model makes it clear that 'for effective WDM policies to be introduced, perceptions of water need to be understood and managed, because such perceptions can mitigate against strategies, which may be well thought out, but which ultimately fail' (Turton 1999:28).

Coping Strategies Versus Adaptive Capacity

A clear distinction needs to be maintained between the concept of adaptive capacity as understood by Turton and the notion of coping strategy (Mohieldeen 1999:23). While adaptive capacity refers to policies, actions and measures that support, and result in, natural resource reconstruction, coping strategies are defined as responses of individual actors or social entities to cope with increasing water scarcity.

Given the context and circumstances coping strategies can or cannot assist natural resource reconstruction (Turton and Ohlsson 1999:7). Responses may include the adoption of WDM principles but can also range from groundwater mining to rainwater harvesting, from political opposition to migration (Turton 1999:3,7).

As with adaptive capacity coping strategies largely depend on the social, economic and political context. It will be shown that a politicised environment, such as is prevalent in the Sa'dah's basin, allows for a range of coping strategies with respect to water scarcity. However, while some of these coping strategies suggest the presence of adaptive capacity others indicate that a politicised environment diminishes the adaptive capacity of actors and social entities.

The extent to which Sa'dah's politicised environment constraints or fosters adaptive capacity will become evident by looking at concepts and notions such as virtual water, food security and food self-sufficiency, the Karshenas' model linking socio-economic development with natural resource reconstruction, values of water, resource capture, belief systems and windows of opportunity.

Virtual Water

The concept of virtual water is useful as it helps to understand the coping strategies during the late 1970s when migration was at its peak. Virtual water is a concept which identifies water, which is embedded in food that enters a country through trade (Allan 1999*b*:2, EC 1998:223). It has been argued that the concept of virtual water is powerful for a number of reasons. First, virtual water provides a remedy for countries facing water deficits. It takes about 1000 tonnes of water to grow one tonne of wheat. Therefore, for each tonne of wheat imported a country effectively saves the 1000 tonnes of water it would take to grow the cereals at home (**Figure 2.2**). The benefits are particularly great in arid environments where agriculture relies on the often unsustainable mobilisation of groundwater resources. Secondly, the option of virtual water allows governments and politicians to downplay or even deny the fact that there is a water deficit and a water crisis (Turton and Ohlssson 1999:19). Therefore, virtual water is a useful coping strategy, allowing governments to postpone the implementation of WDM principles, a process which is likely to be politically stressful and socially disruptive (Turton and Ohlsson 1999:19).

The flooding of Yemeni markets with virtual water, often as free food aid (**Figure 2.3**), during the late 1970s (Morris 1986:138f; 172), allowed for alternative livelihoods to be pursued. Work opportunities in nearby Saudi Arabia resulted in a shortage of the agricultural labour force. In nearby Razih, labour costs during that period increased dramatically (Weir 1985*a* and 1985b:76). Cereal production in Yemen is labour-intensive (Weir 1985b:76) and given the abundance of cheap wheat imports many farmers during that period shifted from the production of cereals to cash crops, especially qat (Swanson 1979:41).

Figure 2.2   Cereal market at Suq al-Talh
The green sacks are barley from Saudi Arabia, the white sacks are
imported wheat flour, and the brown sacks imported wheat (*burr*).

In the context of Sa'dah's politicised environment the abundance of cheap virtual
water created other interesting coping strategies. It is an open secret that free and/or
cheap subsidised wheat imports found their way across the border to Saudi Arabia
where the cereals could be sold with a small profit (Yemen Times 21 August 1995;
Mitchell 1995:4). One Yemeni newspaper even called it 'exporting water' (Yemen
Times 5 June 1995). The import, control and distribution of wheat has became
associated with powerful elites linked to the government and the military (Mitchell
1995:12). Certainly among the rural populations of the Sa'dah basin discussions
about wheat imports bring up images of corruption. For example, profits from the
wheat importing business have enabled a tribal shaykh to donate over YR 50 million
for the support of various charities and development projects, enough money to
establish an educational college and to build, furnish and equip two private hospitals
(Yemen Times 8 January 1996).

The idea of virtual water works effectively in other ways in Sa'dah's political
economy. Cheap subsidised barley from sources in Saudi Arabia is smuggled across
into the Sa'dah basin where it is used to raise livestock, much of which is than
smuggled back into Saudi Arabia for profit. However, the availability of this
particular type of virtual water in the form of barley does not result in water savings.

This is because livestock need additional alfalfa, a high consumer of groundwater in the study area.

Figure 2.3   Virtual water in the form of imported wheat flour
The 35 imported sacks of 50 kg each as seen in the photo represent a minimum of 1600 cubic metres of potential local water saved and freed for higher-value uses such as the domestic and/or industrial sectors.

Virtual Water and the Values of Local Cereal Varieties

In the context of Sa'dah's politicised environment, virtual water as a remedy for water scarce environments faces other socio-economic limitations. This is, first, because local varieties of sorghum are highly valued for social, political and economic reasons. Sorghum is grown by 90 percent of Sa'dah's farmers and many allocate up to 30 percent of their land to it. First, and perhaps most important, sorghum has a number of uses for the rearing of livestock. The long stem, the leaves and the head all provide fodder, which is highly valued. Secondly, sorghum bread is

Figure 2.4   Wheat being stored in barrels for future use
In this case a family was able to get hold of a larger quantity of good quality Australian wheat during election time.

highly valued and an essential part of almost every meal. Thirdly, people in the Sa'dah area have developed such distinct tastes for cereals, including wheat, that they can easily differentiate between the quality of imported wheat varieties. Local perceptions are that much of the wheat imported by the government is of inferior quality, except during times of political election campaigns when superior wheat imports are said to flood the tribal areas (**Figure 2.4**). Local people see these sudden supplies of high-quality wheat as clever political tactics by the ruling elites to buy legitimacy. Whether such 'tasty' imports of virtual water can influence the voting outcome is unclear. They do, however, provide families and communities with coping strategies and help them to stock-up on their food reserves. During one such election campaign, for instance, a tribal family critical of the government was able to obtain a whole truckload of a sought-after Australian wheat variety. At the time of the visit most male members of the family were busy storing the wheat safely in barrels for times of uncertainty.

Food Security

This raises the issue of food security. A distinction needs to be made between food security and food self-sufficiency. A country or political entity is food secure if it has the means to secure the food it requires, often through a combination of growing food and through accessing food available in the global trading system (Lundqvist and Gleick 1997:22). In contrast, the concept of food self-sufficiency, referred to in some sources as 'food self-reliance' implies that a country or political entity has the capacity to produce enough food for its own needs (Lundqvist and Gleick 1997:22).

The oil crisis in the 1970s sparked concerns about food security for many Middle Eastern countries. As a result of the oil embargo the Saudis, in particular, became aware of their vulnerability to a retaliatory grain embargo (Postel 1999:77). Consequently, food self-sufficiency became agricultural policy, although at a high price – '[n]o country has a more dramatic history of groundwater depletion than Saudi Arabia' (Postel 1999:77). Farmers benefited from heavy subsidies for land, equipment and irrigation water. Moreover, their crops were bought by the government at several times the world market price. Saudi Arabia not only reached it's objective of food self-sufficiency but, for a short time, became one of the world's wheat exporters. At the peak of this activity 85 percent of the country's total demand of 20 billion cubic metres of water came from its non-renewable groundwater resources (Postel 1999:77).

The developments just described relate in a number of ways to the factors affecting water management practices and coping strategies in the Sa'dah basin. First, Saudi agricultural policy and its groundwater resources have subsidised livestock rearing in the Sa'dah basin, through a politicised environment, which makes possible the unofficial cross-border trade in cereals, especially barley. Secondly, Saudi Arabia's groundwater 'miracle' created a new awareness of groundwater availability and groundwater lifting technologies among migrant workers from the Sa'dah basin, many of whom worked on Saudi farms.

The concept of food security raises important issues with relevance to the study area. First, there is the question whether the world's agricultural systems can keep up with population growth and meet global food security (Lundqvist and Gleick 1997:22). Postel (1999:6–12,131f) raises a number of concerns with regards to irrigated agriculture, problems relating to salinity in particular. Secondly, increasing demand, triggered by rapidly growing populations, could result in price rises for basic foods, such as cereals. Whether two of the world's largest populations, China and India will be able to remain food self-sufficient or whether these countries will become food importers in the near or distant future has very significant implications, especially for the poorer economies of the Middle East (Allan 2000). In this regard Allan (2001) notes that economies of the Middle East have, over the past two decades, benefited from importing cereals at a price well below its production cost. However, their future import bill for staple foods could rise well beyond their financial competence (Allan 2001). Thirdly, the comparative advantage of economies with the capacity to grow food surplus raises issues of power and political control. As Lundqvist and Gleick (1997:23) put it, '[c]oncerns about the risks of relying on foreign trading partners who may impose conditions on trade or food embargoes for political reasons must first be satisfactorily resolved.' Similar perceptions have also been gained by FAO officials who have observed that '[t]he advantages and risks of relying on international trade to ensure food security are the key issues in discussions concerning alternative food strategies' (Wulf Kluhn of FAO, in Lundqvist and Gleick 1997:22). Rising food import bills and political concerns regarding food security encourage the 'view that countries must be responsible for their own food production...[which] hinders rational solutions to the problem of true food security' (Lundqvist and Gleick 1997:22).

Food security requires, first, for arrangements to ensure that international markets for food are safe and stable, secondly, agreements to minimise the possibility for trade embargoes, and thirdly the capacity of a country 'to improve their economic ability to participate in international food markets' (Lundqvist and Gleick 1997:22). The importance of this last point has repeatedly been stressed by Allan and Karshenas (1996) who make the point that a strong economy increases the options for successful WDM strategies, provides alternatives to supply management, and can lead to what is termed natural reconstruction and a sustainable use of natural resources.

Food Security and Sa'dah's Politicised Environment

Within the context of Sa'dah's tribal society and its politicised environment the idea of food security versus food self-sufficiency raises important concerns among most tribal farmers. Autonomy and food self-sufficiency are important notions of tribal self-definition (Caton 1990a:32). Allan's observation (1999:8) that 'the idea of food insecurity is not part of the 'sanctioned discourse' in most Middle East countries and therefore the relationship between water and food deficit cannot be debated' is certainly true for the tribal region of Sa'dah. The dominant perception among Sa'dah's farmers is that they must be food self-sufficient. Tribal areas have a history of storing grains in hidden cisterns in the ground (Kopp 1977:11) for many years.[1] Even at present, most families in the Sa'dah basin keep grain, tightly stored in barrels, for times of drought, tribal conflict and/or political instability. Indeed, as will become apparent in subsequent chapters, Sa'dah's politicised environment fosters tribal notions of autonomy and food self-sufficiency. Yemen's various crises during the 1990s (1990/91 Gulf Crisis, 1994 civil war, the Hunaysh dispute and continuous tensions along the Saudi-Yemeni border) have led to a reassertion of these values.

The conflict with Eritrea over the Hunaysh islands in the Red Sea, for example, led to fears that Yemen's main northern port of Hudayda could be targeted, putting a stop to imported cereals. Another factor affecting food self-sufficiency relates to tribal perceptions of the state and, in particular, of those involved in the import of food. Sa'dah's tribal farmers do not want to be dependant on, nor controlled by those whom they perceive as corrupt. They are quite aware that food can be used as a political tool by the state to administer control over the northern tribal areas. Interestingly, food self-sufficiency is less of an issue with the area's influential shaykhs and traders. Their political links with the state and the military ensure their 'food security' in times of need.

Karshenas' Concept: the Link between Socio-economic Development and Natural Resource Reconstruction

The theoretical model developed by Allan and Karshenas (1996) is relevant for the study area in that it highlights the relationship between socio-economic development and environmental resource use. The authors argue that a country's environmental capital, such as water, tends to be degraded during the earlier phases of economic development. At a later stage, and depending on the strength and diversity of an economy, a government or community may well have the options and opportunities for more 'sustainable' uses of their natural resources leading to what is termed natural resource reconstruction i.e. the gradual rehabilitation of the environment (Allan and Kashenas 1996).

Figure 2.5a illustrates the concept. The development trajectory passes from sustainable to unsustainable on course for an ecological and Malthusian catastrophe to a point where environmental degradation becomes irreversible.

Two quite different trajectories are shown in **Figure 2.5b**. The lower 'common' or 'conventional' development trajectory describes an earlier phase of development which results in the stock of environmental resources being depleted. Sustainable development, in contrast takes place only along the 'precautionary' development

2.5a) The Concept

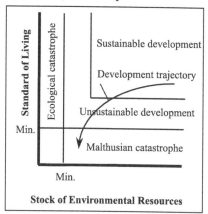

Eco-environmental space:
consequences of low environmental
capacity and 'over-use' of
environmental capital

2.5b) Policies & Practices

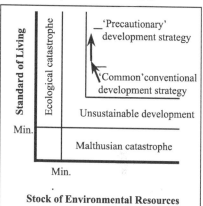

Common & precautionary
development trajectories

2.5c) A Model

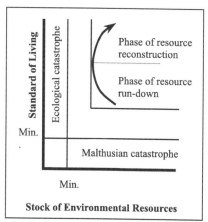

Diversifying economy trajectory

2.5d) Some Empirical Evidence

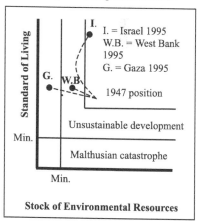

The cases of water use in Israel, West Bank
& Gaza. Trajectories 1947-1995

Sources: Figures 2.5a, b, and c Karshenas, 1994
Figure 2.5d Allan, 1996 (indicative trajectories)

Figure 2.5   Environmental capital (water) and economic development

trajectory, seen here as pointing straight up. Ecologists, especially, argue that development defined as 'sustainable' should not reduce the stock or value of environmental capital for the use of future generations.

The trajectory in **Figure 2.5c** reflects, according to Karshenas, the pattern of diversifying economies (Allan and Karshenas 1996:127). During the earlier phase of economic development a community or state tends to over-exploit its stock of environmental capital. However, the later stages of a strong and diversified economy allow for development options which less rely on the exploitation of limited and scarce natural resources. As a result the phase described as 'run-down of natural resources' can give way to a phase described as resource reconstruction (Allan and Karshenas 1996).

In **Figure 2.5d** Allan and Karshenas (1996) provide empirical evidence from several Middle Eastern economies in support of the theory. A strong and diversified economy enabled Israel in the mid-1980s to adopt economically and environmentally logical policy priorities with respect to water, namely the reduction of water use through cutting water to the agricultural sector. In contrast, the weaker and less diversified economies of Gaza and the West Bank have not provided their communities with the same options. Recently, the concept has also been applied to southern Africa (Turton 1999:5–7).

A number of criticisms have been raised with regard to the Karshenas model. Turton (1999:8) remarks that the Karshenas model 'fails to explain why societies are set on the development trajectory that they are' and suggests that applying Ohlsson's notion of adaptive capacity could provide answers. Handley (1999:57) observes that the model does not address the issue of equity, a point also made by political ecologists, who argue that the economies of the 'north' countries may well enjoy a 'reconstruction of their natural resources at the expense of 'weaker' economies in the 'south', whose natural resources often form the backbone for the continuing prosperity of the former (Bryant 1997:107–121). Another point for critique of the model is the assumption that economic development and environmental reconstruction are complementary (Handley 2001:56). Moreover, Handley (2001:56) points to the difficulty of understanding and predicting the time and ability of a resource to renew itself.

Applying the Karshenas model to the Sa'dah basin highlights the constraints operating in a politicised environment. The strength of the argument made by Allan and Karshenas rests, in part, on the economic concept known as returns to water. Returns to water depend on water use efficiency. Typically, returns to water from agricultural production are modest compared with the returns to water from industrial uses. Water use efficiency can be achieved by implementing measures of 'allocative' and 'productive' efficiency. Allocative efficiency aims to increase returns to water by shifting water resources to uses bringing higher returns. This can be achieved through intra-sectoral allocation – re-allocating water within a sector to activities that achieve higher returns or through inter-sectoral re-allocation – shifting water resources from one sector to another. In water scarce societies such sectoral water efficiency becomes highly relevant as each unit of water creates more livelihoods – 'more jobs per drop' (Allan 1999*a*:6).

Returns to water can also be increased by implementing measures of productive efficiency. At the farm level this has been captured by the slogan 'more crop per drop' (IIMI 1996). Typically, productive efficiency is concerned with technology and institutions to improve distribution and drainage systems within the agricultural sector. In urban and industrial contexts it includes issues relating to water re-use (Allan 1999:8).

The challenges posed by productive and allocative efficiency are quite different (Lundqvist and Gleick 1997:19) Allocative efficiency suggests economically and environmentally logical policy priorities since it achieves much higher returns to water then productive efficiency. However, allocative measures tend to be politically stressful and socially disruptive affecting the life and patterns of communities. In contrast, measures of productive efficiency are politically feasible, since they do not change people's livelihoods or challenge 'vested interests'. This point is vividly made by Allan (1995) who contends that 'if allocative efficiency is not achieved, it is possible, and even common, to be doing the wrong thing extremely efficiently. It would be much more useful to be doing the right thing, that is with efficiently allocated water, a little badly' (Allan 1995 as quoted in Lundqvist and Gleick 1997:19).

How does the Karshenas model and the concept of returns to water relate to Sa'dah's politicised environment? What are the factors enabling and constraining the application of such economic concepts? First, there is evidence of allocative efficiency in the Sa'dah basin. This is mainly intra-sectoral and not inter-sectoral in nature. Starting in the late 1970s farmers shifted increasingly to cash crops. Factors that facilitated the shift were groundwater exploration, improved infrastructure, increased demand for cash crops and the availability of cheap wheat imports (virtual water). Traditionally, activities that involve the growing and marketing of cash crops have been considered as dishonouring to tribal people (Adra 1985:277; Schweizer 1985:112f; Dresch 1993:305). The shift from subsistence to cash crop agriculture indicates some degree of adaptive capacity. Allocative efficiency is also evident during the 1980s. Up to the 1984 fruit import ban the low price for oranges and apples, smuggled across the nearby border, did not justify the production of fruit in the Sa'dah basin. At the time, there were plenty of alternative livelihoods from cross-border trade and migrant labour. The fruit import ban in 1984 increased the 'returns to water' from this sector and most farmers started to grow oranges and apples. However, returns to water from fruit production have since fallen and many farmers have, where possible, re-allocated water to the production of qat, a mild stimulant chewed by a large section of Yemeni society. Demand for qat across the border, where the substance is banned, and a politicised environment, which facilitates cross-border activity, have provided economic incentives for re-allocation.

Values of water are not only defined in economic terms, which explains some of the constraints for intra-sectoral re-allocation. In the Sa'dah basin fruit trees have a high social value. It would be inconceivable that farmers stop irrigating the trees they have tended and nurtured for years. 'What shall I do, I cannot let them die' was the response of a citrus farmer who could no longer afford the water to irrigate his citrus orchard.

In spite of diminishing returns to water from citrus, most influential shaykhs and traders continue to expand their citrus farms. For these actors 'returns to water' has a political dimension. A large irrigated farm, especially citrus, which stays green all year, does much to enhance their tribal-political image. Fruit gifts are also used to buy political favours. Moreover, Sa'dah's politicised environment ensures that shaykhs and traders have less need to consider purely economic returns to water. Over the past decades the Saudi and Yemeni governments have thought it politically expedient to support these actors. In the study area 'gifts' have also included the drilling of wells and the free provision of fruit trees.

Allocative efficiency often implies the shifting of water resources from agriculture into economic activities that achieve higher returns. According to Karshenas' model this process can mark the point where natural reconstruction can begin. With 99 percent of Sa'dah's water resources being monopolised by the agricultural sector the issue of inter-sectoral efficiency requires urgency, especially when considering that groundwater levels are dropping by an alarming rate of 4–6 metres per annum (DHV 1992:44; Fig.4.20). However, inter sectoral re allocation is constrained by Sa'dah's politicised environment. The Sa'dah basin is home to some of Yemen's wealthier traders and businessmen with the capital to invest in economic development. Paradoxically, however, the very factors within Sa'dah's politicised environment that helped many of these trading families on their road to financial power are the same ones which continue to inhibit economic investment (Lichtenthäler 1999:3). Risk factors for economic development include the geopolitics and tribal tensions. As a result, Sa'dah's younger generation is driven to farming for lack of alternative livelihoods.

Resource Capture

Social and political values of land, and linked with that, perceptions of resource capture are other factors which inhibit inter-sectoral reallocation. Resource capture describes the process by which powerful actors and/or social groups seek to gain possession and control over natural resources (Turton 1999:11). Perceptions of resource capture undermine efforts to introduce WDM strategies by politicising a natural resource, such as water (Lichtenthäler and Turton 1999:3). This process is, in particular, damaging, where resource capture is pursued by actors linked to the state, as it destroys the essential elements needed for co-operation, willingness, trust and legitimacy. In other words, resource capture undermines the level of a community's adaptive capacity. If allowed to go unchecked, resource capture acts as a vicious circle, leading to more resource capture. As will become apparent, influential and powerful actors with the ability to chase the falling groundwater table can benefit from the inability of many others who cannot do the same. First, more and more farmers are forced to sell parts of their land to actors who, for political and social reasons, seek to 'capture' these resources. As a result, the latter acquire more and more land, and with it, the right to access and utilise more and more of the area's groundwater resources. Secondly, unlike the average farmer, these influential landowners are able to drill deeper and can afford to install submersible pumps, leading to further resource capture. This process is already forcing a number of poorer farmers in the Sa'dah basin to abandon irrigated agriculture. If the process continues it could reach a point where a kind of 'forced' natural reconstruction begins as control over land and groundwater resources shifts to fewer people.

The process and consequences of resource capture have also been described in the Indian context where 'groundwater overpumping is widening the income gap between rich and poor in some areas. As the water table drops, farmers must drill deeper wells and buy larger pumps to lift the water to the surface. The poor cannot afford these technologies. In parts of Punjab and Haryana, for example, wealthier farmers have installed submersible tubewells at a cost of about 125,000 rupees ($US2,950). As the shallower wells dry up, some of the small-scale, poorer farmers end up renting their land to the larger well owners and becoming labourers on these larger farms' (Postel, 1999:75). A study of the Scottish 'Highland Clearances' (Prebble, 1969) at the end of the 18th century provides evidence that the experience of resource capture is not unique to communities and groups in the south countries.

Figure 2.6 places the Karshenas model within the Sa'dah context. Trajectory F shows the average farmer without additional means of income from non-agricultural activities. He is fast depleting the stock of natural resources (groundwater) within his reach. Shifting to cash crops, especially qat and citrus, in the 1980s, explains the initial horizontal course of the trajectory. However, diminishing returns over the last decade as a result of high inflation and lack of markets has altered the course of his trajectory. Impoverished now, groundwater levels will soon be beyond his reach. Trajectory F+ indicates a farmer with additional income from non-farming activities, either from trade or other entrepreneurial ventures. Such incomes subsidise his farming activities, which are often continued because of the social values attached to farming and home-grown foods. A lack of alternative investment opportunities in the Sa'dah basin also explains his constraints to diversify away from agriculture. If agricultural activities do not create sustainable livelihoods they nevertheless maintain identities and create work. Trajectory S/T indicates influential actors, usually shaykhs and traders with political connections. Income from a number of economic and political sources allows them to abstract the groundwater they need for irrigating their large farms. The social and political values of land and water partly explain why their trajectory has continued on a slightly upward trend. Importantly, at the point where many of the poorer farmers can no longer access and deplete groundwater further, the remaining stock of environmental capital is left for the actors of this socially and politically powerful group. Depending on the 'coping strategies' of the poorer sections of society some degree of 'natural resource reconstruction' could result but at the high price; social dislocation, migration and ecological marginalisation are likely results (Lichtenthäler and Turton 1999:2). In general, however, as Turton (1999:39 n. 40) maintains, 'where resource capture is actively being pursued, natural resource reconstruction is unlikely to occur because it reduces the overall legitimacy of the political system.' He concludes that successful WDM strategies are best served by 'a legitimate system of government with a normative basis that is rooted in the notion of equity' (Turton 1999:39 n.40).

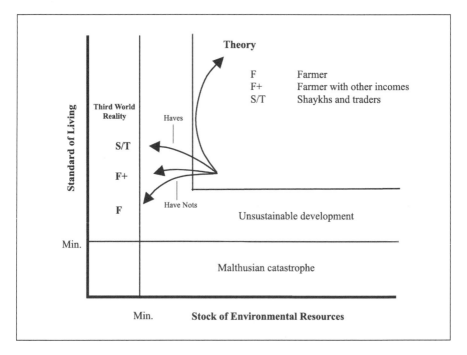

Figure 2.6   Some empirical evidence from the Sa'dah basin

Responses to Resource Capture

The above links to the question of how communities cope with resource capture and the relationship between water scarcity and social stability (Turton and Ohlsson 1999). According to Allan (1999b:10) belief systems provide some answers. He notes that 'belief systems can reflect the experience of five millennia of occasional drought but not of systematic permanent deficit' (1999b:10). Periods of occasional drought lasting for several years can be remembered by older farmers in the Sa'dah basin. During the 1980s, for instance, a drought, lasting several years, forced many semi-nomadic families from the east of Sa'dah to migrate into the basin where groundwater provided secure supplies. Such periods can provide a 'window of opportunity' as it shifts public perception from supply options to demand management, as has been shown in the context of southern Africa (Turton 1999:39 n.38). Good seasons of rainfall, in turn, can result in the public perception that the crisis is over. In the Sa'dah area drought is often linked to a lapse in religious belief and practice whereas rainfall is seen as a religious reward.[2] 'The rains used to be much better' say many of the older farmers, believing that a return to 'normal' rainfall would bring groundwater levels back up. As a result of exceptional rainfall during the dry season in October and November 1997 public perception of Sa'dah farmers shifted from fear and concern to hope and optimism.

In addition, religious beliefs help to explain why resource capture has not resulted in tribal conflict. Inequality, in terms of access to and control over groundwater, is accepted by a large majority of farmers on the grounds of the following two Qur'anic verses (Lichtenthäler 1999:19).

And we raise some of them above others in ranks,
(*Al-Zukhruf*) Surat 43:32 (Translation by Yousuf Ali)

wa Allah faddala ba'dakum 'ala ba'din fi al-rizq
With respect to sustenance Allah has favoured some over others
(*Al-Nahl*) Surat 16:71, translation by Lichtenthäler)

The belief that 'everything is ordered by God' and that the Almighty can change the balance in an instant explains local perceptions. It also puts a majority of Sa'dah's farmers into the 'fatalists' camp (Hoekstra 1998; Thompson 1995:32,33).

The belief that man ought to accept as God-given his state of affairs is by no means unique to Islam as the following text of a well-known 19th century Christian hymn indicates.

The rich man in his castle, the poor man at his gate,
God made them high and lowly, and ordered their estate[3]

Periods of religious renewal can provide 'windows of opportunity' (Kingdon 1984) and the various Islamic schools and interpretations leave room for different interpretations regarding water resources management (Maktari, 1971:3; Vincent 1990:16). The use of religious sermons for raising awareness and to change public perception about water has been demonstrated in some Middle Eastern contexts (IDRC 1998). The flexibility and openness of Zaydi scholarship to 'independent interpretation' (*ijtihad*) has been argued by a prominent Yemeni scholar (al-Amri

1993:190; al-Amri 1982:127). This could present 'windows of opportunity', especially in the light of the present revival of Zaydi belief and practice in the Sa'dah area (Haykel 1995:20f; Lichtenthäler 1999:19). The Islamic principle of *maslahah ammah* (public interest), recognised by all Islamic schools, also suggests new approaches to ownership and abstraction of groundwater (Lichtenthäler 1999:20). Present interpretations appear to give those who own a well on their own land unlimited rights to abstract water for agriculture (Caponera 1973:17).

Land and Territoriality

Beliefs and notions about territoriality and land ownership and rights have also led to perceptions of resource capture, which in turn can trigger a set of responses and coping strategies. Tribes define themselves in term of their tribal territory, which they control and defend. Dresch (1993:334) indicates the 'essential link between territoriality and joint responsibility. If a section or tribe can be held answerable, the territory must be theirs; if the territory is theirs they must be answerable.' This has a number of important implications as seen in the context of the Sa'dah study. First, tribes are reluctant to accept the territorial sovereignty of the state. They feel collectively responsible for what happens within their tribal boundaries. Chapter 4 indicates how these perceptions of territoriality affected the building of a new road to the Sa'dah basin during the mid-1970s. One tribal section refused to have the road run through their territory because they did not want to be held answerable for violations within their boundaries.

Secondly, Dresch's statement indicates that territoriality is questioned in cases where no tribal section accepts responsibility for violations or incidents committed in a given locality. With regards to common land, especially, 'the relations of collective identity to people's rights of use, rights of access, and responsibility are complex' giving rise to ambiguity which can lead to tension and conflict (Dresch 1993:336). This is exacerbated by the fact that in the tribal areas of northern Yemen tribal 'borders are seldom where one expects them' and rarely correspond with drainage systems (Dresch 1993:336). As the Sa'dah case indicates unclear boundaries and contention over territorial claims can provide the state with the pretext to acquire common land resources.[4]

In the study area and until the mid-1970s the vast majority of tribal land was 'white land' (*ard bayda*). 'White land' refers to land not used for agricultural production. Most 'white land' was communally owned and used as grazing areas for the respective communities. In turn, the grazing areas provided essential runoff for small privately owned agricultural plots further down stream. The complex relationship between runoff rights and 'white land' effectively put a stop to the development of new land. This is confirmed by Dresch (1993:359) who notes that '[t]o bring new land into production or sink a new well requires permission from all those who are connected to the site by shared runoff, which in settled areas means that no permission is forthcoming.'

1. Runoff area south of al-Dumayd.

2. Runoff area south of Suq al-Talh.

3 Runoff area in Bani Mu'adh

5 km   White areas indicate land not used for agriculture.

Figure 2.7   Selected runoff zones in the Sa'dah basin

1. Runoff zone south of the water stressed area of al-Dumayd where rainwater harvesting is preferred over groundwater-irrigated agriculture due to the increasing awareness that groundwater levels in the al-Dumayd area are fast dropping beyond the reach of all but a few.
2. Runoff zone south of Suq al-Talh and near the village of Al Abin, which has largely escaped groundwater exploitation because of tribal dispute over claims to runoff.
3. Runoff zone in Bani Mu'adh. As a result of tension between Bani Mu'adh (Sahar) and tribal groups from Hamdan Bakil this large mahjar has been stopped from groundwater development.
   Date of satellite image: August 1998.

'Runoff rights are stronger than rights to land' (*al-sabb aqwa min al-salab*). From the late-1970s onwards claims to runoff facilitated the process of resource capture as Chapter 4 will show in more detail. In brief, during the 1970s surface water runoff was essential for agricultural production in some areas and less significant in others. The financial and technological possibilities to abstract groundwater gave runoff rights a new significance. Following arbitration in 1972, runoff rights were valued at half of the total runoff area. This process solved some disputes over scarce surface water resources but created many new conflicts over claims to runoff. Being able to establish such claims provided enormous potential for resource capture. Resulting conflicts, some of which are ongoing, explain many of the 'white' (agriculturally unused) lands left in the Sa'dah basin, as seen in the satellite image (**Figure 2.7**). It is likely that some of these areas will escape future groundwater exploitation. During the 1970s and 1980s runoff rights helped to 'capture' groundwater resources, through the acquisition of land rights. With the rapid fall of the groundwater table, remaining runoff rights are increasingly valued again to secure adequate surface water supplies in the future.

Islam and the Notion of ihya al-mawat (bringing new land into production)

Islamic interpretations of land development have fuelled perceptions of resource capture by the state. This in turn has triggered tribal responses to 'secure' their own resources. According to Islamic jurists *mawat* (dead land) refers to land neither owned nor used. Bringing 'dead' land into production (*ihya al-mawat*, revival of *mawat*) is desirable and should be encouraged. To achieve this most jurists 'favoured a system of profit motive by which developers of *mawat* land become legal owners of that land as soon as the intended development is complete (Maktari 1971:9). The Hanafi school added an interesting clause – the Imam (ruler) had to give his consent to the acquisition of *mawat* land. Rulers, in south-west Arabia in particular, have preferred the Hanafi interpretation because it gave them considerable control over waste lands. Maktari (1971:10) relates one example from Yemen which shows how rulers and governments have applied the notion of 'revival of land' to claim possession of tribal land. In one particular case the government introduced a law for the lease of 'waste' land. This move provoked strong reaction from the local tribe who 'claimed that the land in question lay within their tribal boundary, and that they did not recognise the right of the government to lease their land. They claimed that the desert surrounding their villages, including all wells, was theirs in perpetuity' (Maktari 1971:10). The story indicates that tribal perceptions of territoriality and land classification contrast with those who want to come into the possession of 'dead' land. It also becomes clear that states and rulers can use interpretations of 'revival of land' to enlarge their sphere of influence.

Maktari (1971:10) maintains that by applying the principle of 'revival of land' (*ihya al-mawat*) sultans and rulers in south-west Arabia 'bestow upon their supporters gifts of land' and 'some sultans have become accustomed to marking off certain *mawat* land as their own property and that they maintain for themselves the right to charge a rent to those who use it.' Furthermore, Islamic law stipulates that '*mawat* land may be declared to be state property reserved for public use' (ibid).

Up to the late 1970s much of the Sa'dah basin was land not agriculturally used (*ard bayda*). As discussed in Chapter 4, tribal sections were concerned about the application of 'revival of land' (*ihya al-mawat*) to some of their tribal wastelands. Interestingly, the way to safeguard tribal land from acquisition by the state or the military was itself spelt out in the Islamic interpretations about *ihya al-mawat*. It is worth quoting the passage by Maktari (1971:10):

> [Shafi'i law] 'regards ownership of mawat land as complete and binding from the moment an intending developer takes full possession. It defines the condition of full possession as the actual and complete development of the land for the purpose intended. For example, if the land was intended for building a house, the actual building of the house fulfils the condition of possession. Or if the land was intended for cultivation, mere preparation, which should include providing water for irrigation, suffices.'

The passage helps to understand the reactions of individuals and communities in the Sa'dah basin. First, ownership of *mawat* land could be secured by the mere process of developing land. The building of a house or the mere provision of a well provided sufficient proof of ownership. Both of these steps were applied by communities and farmers in the Sa'dah basin to secure their 'white' lands. In some areas farmers were told that the government would take their 'white' lands unless they themselves 'revived the land'. In other cases mere rumours about *ihya al-mawat* triggered the preparation of new farms. State-sponsored hydro-geological surveys of the area, during the early 1980s also led some farmers to drill wells and prepare fields for cultivation.

Islamic water rights pose other constraints on adaptive capacity and can aid resource capture. Water from a well on privately owned land becomes the property of the owner (Maktari 1971:15, Caponera 1973:17). In times of need and shortage customary obligation demands that privately owned water be shared to satisfy the thirst of people and animals. However, the owner of a well has no legal obligation to share the water for the irrigation of land (Maktari 1971:19,25). Ownership of land, therefore, gives access to groundwater resources, the 'capture' of which can be justified by Islamic law.

Adaptive capacity is further constrained by 'legal pluralism'. Contradictions with respect to the ownership of groundwater exist between Yemen's constitution and its civil law. Constitutional Article 8 makes the state the legal owner of groundwater 'All types of natural resources whether above ground, or below... are property of the State.' (Haddash 1996:5). In contrast, Yemen's Civil law, Article 1366, which is based on Islamic *shari'ah*, maintains that 'water is originally *res nullius* for all (*mubah*)...'. Applying the latter, Haddash points out, 'an aquifer is freely accessible, with no control, to anyone who owns the overlying land and can afford the cost of drilling a well' (1996:5).

Natural Resources Managed in Common or a Tragedy of the Commons?

The question whether this situation supports the 'tragedy of the commons' theory is explored by Kohler (1999:4) who argues that Yemen's institutional framework has not kept pace with modern technological possibilities of abstracting groundwater.

The 'game' theory of the 'tragedy of the commons' maintains that natural resources are likely to be overexploited if property rights are communal. The arguments imply that overexploitation of natural resources is especially a problem when there is competition over resources and communities cannot intensify the use of a resource further.

To avoid what has been termed the 'tragedy of the commons' Hardin (1968) suggests that one way to go forward is to rearrange the ownership of the resource i.e. to privatise natural resources managed in common. Game theory argues that destruction of the resource will be the natural outcome if the ownership of the resource is not defined.

However, some have argued for a distinction to made between 'restricted access resources' and 'open access resources' (Ostrom 1990:48). Moreover, whether the management of a common property resource will result in a 'tragedy of the commons' is seen as matter of scale. There is empirical evidence to suggest that smaller homogeneous communities that are heavily dependent on their natural resources for economic returns have the ability to manage common property resources in sustainable ways (Ostrom 1990:26).

Privatisation or state control are advocated as remedies against a 'tragedy of the commons' (see Ostrom 1990:12). However, Allan points out that there are equity implications which are impossible to incorporate into a privatising approach and that political issues will determine who can gain ownership (private communication 22 Dec. 1996).

In the past shallow wells and animal power naturally controlled and limited the volumes of groundwater abstracted in the Sa'dah basin. These factors defined groundwater as a 'restricted access resource'. In addition, the 'restricted access' character of communal tribal land imposed 'restricted access' on groundwater resources. No wells could be drilled where runoff rights were attached to communal land. It was privatisation of communal lands from the mid-1970s on that facilitated a change in status for groundwater. Through privatisation groundwater became an 'open access resource'. It is this shift from a 'restricted access' to an 'open access', which is gradually leading to a 'tragedy of the commons'.

Institutions and Community Participation

A shift to sustainable water management raises questions about the role of the state and the local community. The pivotal role of local communities in managing their water resources is increasingly recognised (World Bank 1997a:1 and 1997b:16). Writing about communities in the Yemeni highlands Vincent (1990:26) argues that '[i]n an environment characterised by small, disparate water supplies, and limited central government involvement in local affairs or production, community management may offer most potential.' She concludes that 'communities can manage their water resources if local institutions have the legitimacy to do so, and if social and economic incentives encourage their performance...'

The current situation in the Sa'dah basin, however, is unlike that in traditional communities with a history of addressing and solving their own water problems. Co-operation over groundwater management is constrained by tribal politics. Since the early 1980s and especially after the fruit import ban in 1984 large areas of grazing land in the Sa'dah basin have been sold off to individuals and families belonging to tribes from outside the Sa'dah basin area. The new landowners included people from the two main tribal confederations Hashid and Bakil with their numerous subsections as well as from the tribes of Khawlan b. Amir with their main subsections Razih, Munabbih, al-Mahadhir, Khawlan and Juma'ah. With the exception of a few individuals these new landowners have not changed tribal affiliation by moving among the Sa'dah tribes. They share no history of co-operation with their host communities. In fact, a primary reason for moving to the Sa'dah basin may have been to break free from the need to share and co-operate over the scarce and limited water and land resources in their highland home territory. Informants argued that these factors presented major constraints for addressing water issues. This is also confirmed by Vincent who notes that water management is likely to pose a problem 'in villages with internal divisions and historic feuds over water resources' (1990:23).

Another central factor, which prevents farmers and communities from addressing the groundwater abstraction issue, is related to the established interpretations in Islamic law. Kohler (1999:138) provides a vivid example for this. When he asked farmers whether they discussed their water problems with up-stream communities they answered that there was nothing to discuss because they had all switched to groundwater irrigation. The same is true in the Sa'dah basin. Everyone accepts as legitimised by Islamic law the right of a landowner to abstract whatever groundwater he can and needs regardless of the impact on other users and the environment.

However, it has also been pointed out that 'Islamic law can be flexible and pragmatic' (Vincent 1990:16). The Islamic principle of *maslahah ammah* provides one 'window of opportunity' because it recognises that the interests and welfare of the wider community have priority over and above individual rights and benefits even if these are lawful (*al-maslahah al-ammah muqaddamah ala al-maslahah al-*

khassah). In support of this principle some scholars argue that 'the *shari'ah* has to be applied wherever the general interest lies' (*haythuma kanat al-maslahah fathama shar' Allah*).[5] Based on the notion of 'no harm' it appears that the concept of *maslahah ammah* can be explored to help regulate groundwater abstraction and well drilling (see also Lichtenthäler and Turton 1999).

Tunisia provides one example where the Islamic principle of *maslahah ammah* has been applied. In central Tunisia farmers are not prevented from digging new wells, but they may have their extraction of water restricted in volume, in the community interest. (Vincent 1990:16).

The recognition in the early 1970s that it was in the overall interest of all tribes to re-negotiate established rights to land and water led to fundamental changes with respect to resource management. As many farmers arc becoming increasingly aware that regulation and groundwater management will indeed serve the long-term interests (*maslahah ammah*) of their respective communities a large majority might again look to respected men of religion to arbitrate new approaches to groundwater management.

For such change to happen the important role of education and information for raising awareness is noted by many authors. To raise awareness will be the key. Lundqvist and Gleick (1997:ix) argue that 'free flow of information and the use of hydrological information systems are prerequisites for proper water management.' The EC guidelines on water resource management issue an even stronger statement by saying that '[t]he most important issue for stakeholders is information' (EC 1998:251). In addition, they advocate that education 'should not be restricted to conventional methods as different types of education will be needed for different situations' (EC 1998:234).

These guidelines clearly 'discourage the perpetuation of centralised and hierarchical command structures for water resource management'. They also recognise the subsidiarity principle; i.e. '[r]esponsibilities for water-related services and resource management need to be decentralised to the lowest appropriate administrative level according to the concept of subsidiarity' (EC 1998:221).

However, for the state and the technocratic elite in Yemen to recognise and accept the role and potential of Sa'dah's communities to address water issues perceptions of the tribes need to change. Weir (1998:1) notes the various stereotype images in currency about Yemeni tribes. The ruling elites of the imamate and the republic have variously portrayed them as 'anarchic, militaristic, irreligious polities, continually feuding and inherently opposed to states...in order to justify their subordination to the state' (Weir 1998:1). Even today, these images linger on as evident from a recent article written by a Professor of Sociology of San'a University (Yemen Times 2 March 1999). He perceives the tribes as a hindrance to Yemen's social, economic and political development. He even accuses them of looking down on agricultural work, a claim that no one with any knowledge about Yemen's highland tribes can maintain (Adra 1985:280; Dresch 1993:307).

The following chapters will reveal the set of dynamic coping strategies available to actors and communities operating in politicised environments such as the Sa'dah basin. However, it will also become evident that a politicised environment constrains the type of adaptive capacity, which allows for natural resource reconstruction to take place. This is because resource capture politicises water thereby undermining water demand management strategies.

Notes

[1] Sergeant (1995 VII p 68 quotes Hamdani, Sifat, 107ff 'I have seen in J. Mawar wheat over which thirty years had passed without it stinking and changing. As for millet it is only in a hot district, and it is not stored in houses on account of the rotten state that soon overcomes it, but excarvations are made for it in the ground and it is buried in silos (madaafin), a single one holding 5,000 qafiiz or less. It is then closed over until perhaps even thorn bushes grow on the cover, and it lasts a lifetime withour being lost (Qafiiz consits of 48 mudds and to vary from 25–55 litres the lower figure in the early Islamic centuries being usual.'

[2] Morris (1986:236f) notes that '[a] relationship is perceived between man's worthiness to receive blessing and the level of precipitation. Rainfall is believed to fall through seven successive heavens and is held in the lowest, just beyond the vision of man, while Allah gauges the morality of those who hope to receive it.' Moreover, he observed that '[s]ome older men seem convinced that the apparent continuous shortage of rain is a judgement on the impious nature and abundance of envy in modern society' (1986:237).

[3] D. Martyn Lloyd-Jones, 'one of the greatest Bible expositors in the English-speaking world' points out the 'grievous harm that has been done to the Church of Christ because a statement such as this ...has been misinterpreted' saying, that 'there is nothing in the Bible that disputes the proposition that all men are equal in the sight of God and that all are entitled to equality of opportunity' (Lloyd-Jones 1981:279).

[4] Territory can be lost to a tribal section for a few other reasons. In case of a village unit changing tribal affiliation their territory remains with them. Territory can also be ceded as payment for blood money, provided that the debt is clearly collective (Dresch 1993:333).

[5] Personal communication with Muhammad Abdul Halim, Prof. for Islamic Studies, SOAS.

3 Environment, Society and Economy of the Sa'dah Basin

We have three problems in Yemen: qat, shaykhs and water
(A Yemeni from San'a)[1]

The aim of this chapter is to introduce relevant aspects of the Sa'dah basin's environment, society and economy (**Figure 3.1**). It will be been shown that, until the mid-1970s, scarcity of rainfall and lack of surface water resources limited and defined agricultural activity. Until the early 1970s, scarcity of water resources shaped and determined not only population density and agricultural settlement patterns but economic activity as well. Intensive agriculture was mainly confined to some wadi areas while rearing livestock was prominent in the basin itself.

In addition this chapter will explain the extent to which Islamic and tribal-customary laws, pertaining to the allocation and management of the Sa'dah basin's scarce surface water, define the relative power of one actor over the environment of another. In this context it will be demonstrated that extensification of groundwater irrigated agriculture was, until a crucial arbitration in 1972, impossible since a) most land was communally owned and managed and b) extensification was blocked by the power and control exercised by down-stream owners of runoff rights. The solution to the impasse mediated by a religious scholar in 1972, which consequently altered the control and balance of power with regard to land and water resources in the Sa'dah basin will be explained. The chapter will then introduce the area's influential actors. Their quest to 'capture' and control the basin's natural resources will be outlined.

Environment

Location and Topography

Figure 3.2 shows the geographical location of the Sa'dah basin while **Figure 3.3** represents a cross-section with the main sub-regions based on Gingrich and Heiss (1986:28) and Gingrich (1993:256).

The Sa'dah basin is situated at the northern part of the central highlands of the Republic of Yemen and about 80 km south of its boundary with the Kingdom of Saudi Arabia. The basin extends 30 km to the NW-SE and, at its widest point, 16 km to the SE direction and covers a total area of 213 km² (DHV 1992:6). Topographic

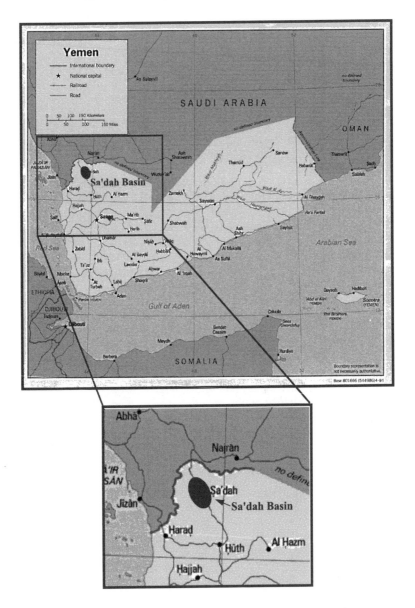

Figure 3.1   Map of Yemen with the location of the Sa'dah basin

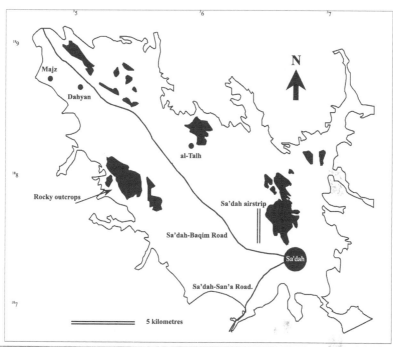

Figure 3.2   Map and photo of the Sa'dah basin
The alluvial plain covers an area of 213 km². The map shows the provincial capital of Sa'dah, the San'a-Sa'dah road entering the town from the south-east and the Sa'dah-Baqim road running north through the basin to the Saudi border.

WEST

EAST

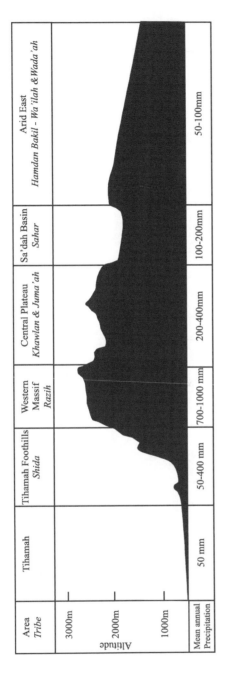

Area *Tribe*	Tihamah	Tihamah Foothills *Shida*	Western Massif *Razih*	Central Plateau *Khawlan & Juma'ah*	Sa'dah Basin *Sahar*	Arid East *Hamdan Bakil - Wa'ilah & Wada'ah*
Altitude						
Mean annual Precipitation	50 mm	50-400 mm	700-1000 mm	200-400mm	100-200mm	50-100mm

3000m 2000m 1000m

Rainfall Sources: Van der Gun & Ahmad (1995)

≡ 20 km

Figure 3.3   Cross-section of the Sa'dah region, from the Red Sea Tihamah lowlands in the West to the desert of the Empty Quarter in the East

elevations range from 2050 m in the NW to 1840 m in the NE giving the basin a gentle slope of 1.2 per cent. At its western edge the basin is surrounded by rugged mountains that rise steeply to elevations of 2700 m (Van der Gun 1985:6). Their escarpment marks the main watershed that crosses the Yemen in a north-south direction. Here, several wadis drain into the Sa'dah basin. By contrast, the mountains defining the Sa'dah basin on the east only rise 150-500 m above the basin. There is a gradual decline in precipitation and wadi flows are less frequent here than in the west.

Climate and Rainfall

According to a UNESCO climatic classification, which is based on the ratio between average annual precipitation and annual reference evaporation, the Sa'dah basin must be described as arid (Van der Gun 1995:36). The high altitude of the basin (1800m) ensures moderate temperatures ranging from 13° C in January – 25°C in July and averaging 19.3°C annually (DHV 1993*b*:13).

Figure 3.4 shows the main rainfall pattern in the Sa'dah basin. Rainfall is sporadic, often comes in short and intense outbursts and can vary greatly between local areas. Precipitation can occur during the whole year but usually peaks during the two main rainy seasons – March-May and July-August. Total annual rainfall is low as indicated by a mean of 129 mm between 1983–1991. It also varies significantly from one year to the next – 271 mm in 1983 and as little as 58 mm 1984. Average relative humidity is only 43 per cent (Van der Gun 1985:12). This, together with the low figures for rainfall explain why evaporation far exceeds natural precipitation. A German-Yemeni Plant Protection Project measured 2800 mm of pan evaporation for 1976. Another study (Van der Gun 1985:12), using the Penman method, calculated the annual potential evapotranspiration as high as 1360 mm.

Prior to the introduction of tube wells in the early 1970s the Sa'dah basin's natural vegetation mainly supported the rearing of livestock (Gingrich and Heiss 1986:15,27). Its savannah-type grasslands have mostly disappeared as a result of intensive grazing. Consequently, non-cultivated areas of the Sa'dah basin give the appearance of a desert (DHV 1993*b*:13).

Surface Water

A number of variables determine the surface water resources of a given area; precipitation patterns, rates of potential evaporation and evapotranspiration, terrain characteristics such as geology and soil, land use and human interaction (Kopp 2000:80). Runoff constitutes '[t]hat part of precipitation which is neither absorbed into the ground, stored on the surface, nor evaporated, but which flows over land.' (Barrow, C. 1999:137). The percentage of precipitation which gets converted into runoff in turn depends on land surface characteristics, such as slope inclination, vegetation, vegetation cover, soils, geological formation, rates of evaporation and infiltration (Kopp (2000:80).

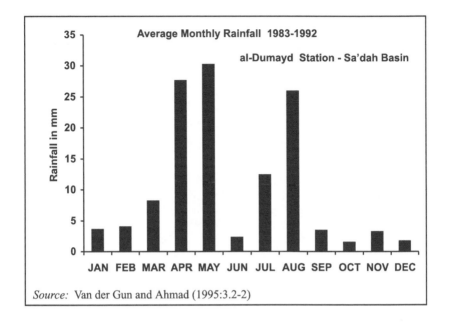

Source: Van der Gun and Ahmad (1995:3.2-2)

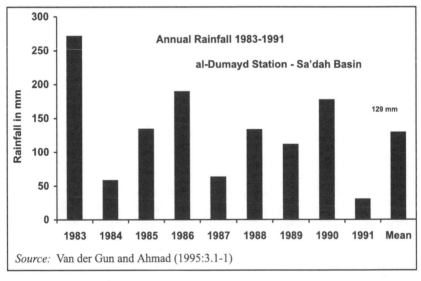

Source: Van der Gun and Ahmad (1995:3.1-1)

Figure 3.4   (A) Average monthly rainfall for the Sa'dah basin between 1983–1992 (above)
(B) Annual rainfall in the Sa'dah basin between 1983–1991 (below)

The Sa'dah basin is part of Yemen's Eastern Escarpment which drains into Arabia's Empty Quarter (Van der Gun and Ahmad 1995: Figure 5.1). However, the basin has no permanent or seasonal streams and only after exceptional heavy rains will surface water runoff be discharged into Wadi Marwan leaving the basin to the north-east from where it subsequently flows into Wadi Najran (Van der Gun 1985:17). For the most part, runoff carried by various wadis entering the basin from the west, south and east quickly and almost entirely infiltrates the permeable alluvial deposits and recharges the underlying aquifer. As a result the Sa'dah basin has been characterised as a 'run-off absorbing zone' (Danikh and Van der Gun 1985:24). These factors as well as the basin's geomorphological conditions account for Sa'dah's considerable groundwater resources.

Groundwater

Groundwater is defined 'as the subsurface water of the saturated zone, i.e. the zone in which the pores and fissures of the soil and rocks are, in principle, completely filled with water' (Van der Gun and Ahmad 1995:64).

Sa'dah's Wajid sandstone aquifer is classified 'as highly productive' (Van der Gun and Ahmad 1995: Fig. 6.1) making the basin a relatively favourable area for groundwater development. This is for the following reasons:

- Groundwater levels are (or used to be during 1982/1983 survey) only 20–40 m.
- The hydraulic conductivity of the surrounding rock units is lower than those in the basin.
- Alluvial deposits and the basin's geomorphological structures favour groundwater recharge.
- Surface waters are mainly contained in the basin. Discharge occurs rarely and in insignificant amounts.

Society

Tribal Divisions and Groups

Most Yemenis refer to the plateau areas north and north-east of the capital San'a as the land of the tribes *bilad al-qaba'il* or the land of Hashid and Bakil, *bilad Hashid wa Bakil*, as the two main tribal confederations who inhabit the area are known. Further north still lies the Sa'dah region or *mintaqat Sa'dah*, a term used to distinguish it from *bilad Hashid wa Bakil* (Gingrich 1993:257). And while a number of Bakil tribes inhabit the eastern arid parts of the Sa'dah region the majority of is tribes belong to neither Hashid nor Bakil but form their own cohesive tribal confederation called Khawlan b. Amir.

The confederation of Khawlan b. Amir is made up of eight tribes as shown in **Figure 3.5**. Since 1934, when Yemen lost territories to Saudi Arabia, three of these Khawlan tribes, Fayfa, Bani Malik and Bani al-Ghazi (Gingrich 1993:258), now live across the present border, in Saudi Arabia. The remaining five which, in terms of numbers, make up the majority of the Sa'dah region are Munabbih, Razih, Khawlan, Juma'ah and Sahar. Munabbih and Razih inhabit isolated mountains towering over the low Tihamah coastal areas in the west. To the east of these two areas is Khawlan which extends eastwards up to the Sa'dah Basin.

The Sa'dah basin itself is largely the home of the two remaining tribes with the Sahar occupying most the basin apart from the northern part which is the home of Juma'ah. Only the basin's eastern and south-eastern peripheries do not belong to the Khawlan confederation but are the tribal territories of two Bakil tribes, Wa'ilah and Wada'ah.

The Sahar, who claim the lion's share of the Sa'dah basin, belong to two main groups, Kulayb and Malik (Gingrich and Heiss 1986:170 n.120), each of which has a number of tribal sub-groups (**Figure 3.6**). Of the five main groups that associate with Malik three own most of the land in the Sa'dah basin. They are Bani Mu'adh, Walad Mas'ud and al-Talh. Malik's remaining two groups have their tribal home south of the basin, al-Mahadhir occupies a small plain just south of the Sa'dah basin while Bani Uwayr command a higher mountain plateau west of al-Mahadhir.

The six tribal sub-sections of Kulayb appear to be placed at the basin's edges where some of the major wadis drain into the Sa'dah basin, al-Abdin and Ghuraz just south of the provincial capital Sa'dah, Wadi Alaf to the south-west, al-Uzqul at the main watershed to west of the basin, Al Dhuriyyah in the rocky outcrops of the northern corner and al-Abqur, a wadi coming in from the north.

Before the unprecedented economic and political changes in the mid-1970s, which facilitated the exploitation of Sa'dah's main aquifer, human settlements tended to be at or near the main Wadi courses. In contrast, the central parts of the Sa'dah basin did not support much agricultural activity as it lacked the necessary surface water resources (Gingrich and Heiss 1986:27; Van der Gun 1985:16). With this consideration in mind it might appear as if Kulayb located themselves on the lands favoured for surface and groundwater resources.

Other tribes living in between Sahar   On the basin's eastern and south-eastern peripheries the Khawlan b. Amir tribes are surrounded by several Bakil tribes (**Figures 3.7, 3.8**). In local terminology their territories are called *bilad Hamdan* (The home of Hamdan). Among them are the Wa'ilah tribe whose tribal lands extend all the way east to the city of Najran, at the edge of Arabia's Empty Quarter. They also control some of the main routes that lead to Saudi Arabia. The territories of Wada'ah fall in the south-east of the basin in upper Wadi al-Abdin while Al Ammar controls the territories along and east of the main road that connects the Sa'dah basin to the capital San'a.

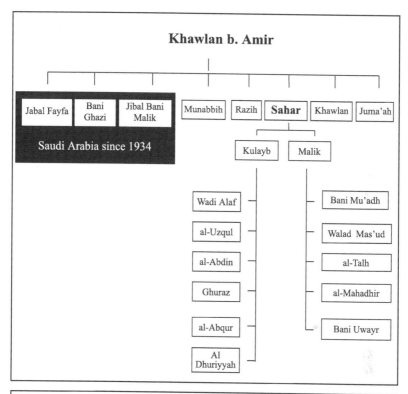

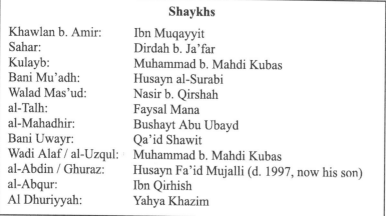

Figure 3.5 **Tribes of Khawlan b. Amir, especially Sahar (Sa'dah basin) with its subsections and shaykhs**

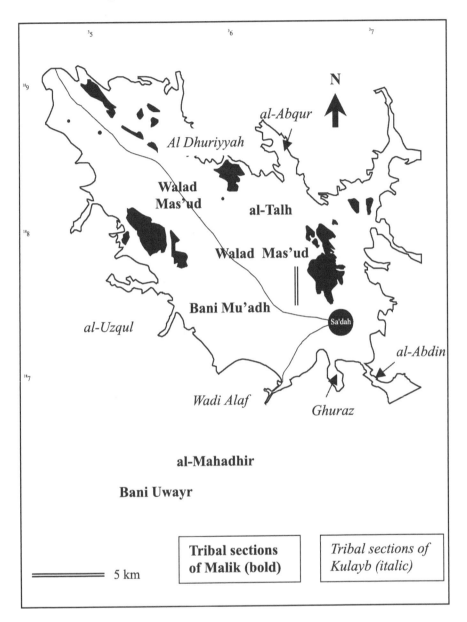

Figure 3.6   **Spatial distribution of Sahar tribal groups in and around the Sa'dah basin**

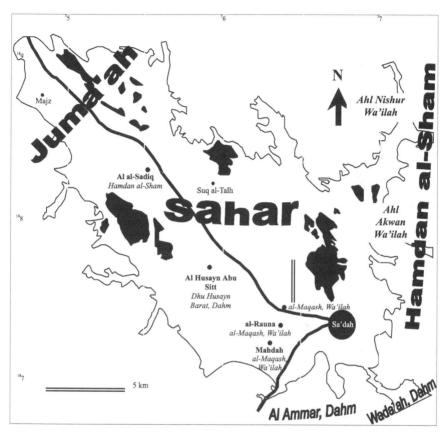

Figure 3.7 Location of tribal groups in Sa'dah area
Juma'ah and Sahar are part of Khawlan b. Amir. At the basin's eastern
peripheries and extending to the desert of the Empty Quarter are the
tribes of Hamdan al-Sham (Bakil).

Since the distant past a few tribal families from Bakil have settled among the
Sahar in the Sa'dah basin. The village and area of al-Maqash, a subsection of
Wa'ilah (Bakil), lies about 5 kms north of Sa'dah town along the main asphalt road
that runs north through the basin. Another Bakil village is Mahdah just two kms
west of al-Maqash. A few kilometres further north along the main road and just next
to Walad Mas'ud's qat market lies the village of Al Abu Sitt (Bakil) which enjoys
close connections with the Dhu Husayn tribes of Barat in the east. And then there is

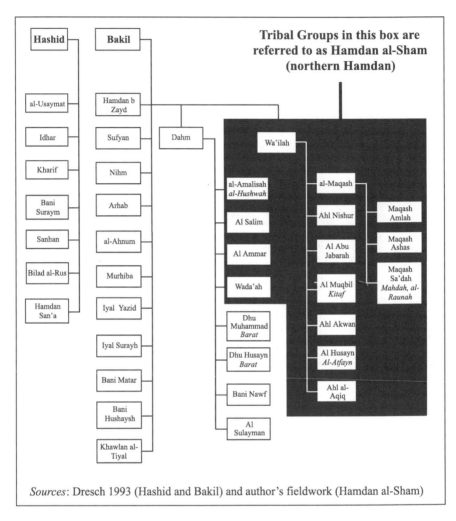

Sources: Dresch 1993 (Hashid and Bakil) and author's fieldwork (Hamdan al-Sham)

Figure 3.8   Tribal groups of Hamdan al-Sham (Bakil)
Some of these belong to Wa'ilah while others are part of Dahm. A number of Dahm groups, however do not belong to the the northern Hamdan, among them the Barat tribes at frequent conflict with one another, Dhu Muhammad and Dhu Husayn. Wa'ilah and Dahm, too, have a history of tribal conflict (see Dresch 1993:347).

the village of Al al-Sadiq (Bakil), which is likewise located in the centre of the basin, just north of Ahma al-Talh and to the west of the famous tribal market of Suq al-Talh. Here shaykh Faysal Mana of al-Talh has his large home. As an elected MP for the al-Talh constituency (Member of the House of Representatives, *majlis al-nuwwab*) he often travels between his home area and the capital and his personal driver happens to be a tribesman from this Bakili village of Al al-Sadiq.

While the tribal concept of *juwarah* (neighbour right) usually tries to prevent the selling of land to members of another tribe, land may sometimes be transferred through settlement of a blood feud (Dresch 1989:81; Kopp 1981:136).[2] The village of al-Maqash, for instance, is said to be protected enclave (*hijrah*) for the eastern Hamdan tribes. However, the unprecedented economic and post civil-war conditions of the 1970s facilitated migration by all kinds of tribal as well as non-tribal people into the Sa'dah basin for reasons that are discussed in Chapter 4.

The Khawlan b. Amir confederation has one paramount tribal leader (shaykh) who can, in theory, rally the Khawlan tribes for united action. But each sub-tribe has a shaykh too. Traditionally, his standing has been measured by his ability to solve tribal problems and disputes. Nowadays his 'weight' and influence is increasingly measured by how much his 'belly' is 'full of politics' (Dresch 1989:100). Since the 1970s many Sa'dah shaykhs have become influential players in regional and national politics. Their large farms are symbolic of their new wealth. But as Dresch (1989:100) remarks, as long as he has the power to settle disputes '[m]en will follow such a shaykh even if he 'eats' their money.'

The overall shaykh of all the Sahar tribes is Dirdah b. Ja'far who resides in Harf b. Ja'far in the territory of the Bani Mu'adh. However, his standing with the Imam and the Royalists during the 1962 revolution subsequently reduced his political power and influence. At the same time those shaykhs that supported the Republican cause during the civil war have gradually gained power and influence in the region.

Since the late-1970s, when the Sa'dah basin became the main trading point for smuggled goods from Saudi Arabia, many Sahar shaykhs of territories outside the actual Sa'dah basin moved in to be close to the emerging centres of tribal political and economic power. Moreover, even shaykhs who are not part of the Khawlan b. Amir group – from the Hamdan tribes of the east were able to establish themselves in the Sa'dah basin. The dynamic processes which made this movement possible are the focus of Chapter 4.

Religious Groups

Besides its tribal population the Sa'dah basin is home to many *sayyid*s (pl. *sadah*). The *sayyid*s are the doctors of religion and claim descent of the Prophet. Since the beginning of the 9th century when al-Hadi ila al-Haqq Yahya b. al-Husayn founded the Zaydi Imamate in Sa'dah, *sayyid*s *(sadah)* have lived in protected enclaves (*hijrah*) within tribal territory. Many Zaydi *hijrah*s are centres of religious scholarship and the *sayyid*s act as arbitrators and religious teachers.

The Sa'dah basin is host to a number of such *hijrahs*. The provincial capital of Sa'dah is the location of al-Hadi's tomb and has always been a stronghold of Zaydism. Other larger *sayyid* settlements are Dahyan, 25 km to the north of Sa'dah town, Hamazat, at the eastern edge of the basin (Heiss 1987:63), and Hijrah Fallah in Wadi Fallah.

Over the centuries certain *sayyid* families have come into the possession of large land holdings through the Islamic institution of *waqf* (religious endowment). A tribal member donates a piece of land to the mosque after which the land becomes the property of the Muslim community (*mal al-muslimin*). The Imam of the mosque then makes a *sayyid* his partner (*sharik*). The *sayyid* either works the land himself or gets someone else as a *mushrik* (partner) to work it in return for part of the harvest crop. In theory the Imam could ask the *sayyid* to return the land; in practice, however, large landholdings have come into the quasi permanent possession of *sayyid* families in this way. The large *waqf* estates around Sa'dah town are a case in point. Gingrich (1986:21) remarks that it was one of the aims of the Revolution to abolish the misuse of this practice.

Economy

Pre-Islamic and Early Islamic Period

Sabaic inscriptions already referred to the town of Sa'dah as part of the territory of the tribe of Khawlan b. Amir (Encyclopaedia of Islam 2 Vol.VIII:705). During pre-Islamic times, and as early as 400 BC Sa'dah was an important settlement along the main trade route that linked the kingdoms of Southern Arabia with other parts of the ancient world (al-Thenayian 1997:243; Gingrich and Heiss 1986:14; Heiss 1987:66; Grohmann 1933:70; Wald 1980:172). After the establishment of Zaydi Islam in 897 AD Sa'dah gained importance as the spiritual and often political capital of the Zaydi imamate in Yemen (Encyclopaedia of Islam 2 Vol. VIII:705; Gingrich and Heiss 1986:15).

During the times of al-Hamdani in the 10th century the area had become renowned for the manufacture of many iron implements, in particular iron arrowheads (*nisal sa'diyyah*) (Encyclopaedia of Islam 2 Vol. VIII:705). Ibn Mujawir, travelling the area during the 13th century, describes Sa'dah as a flourishing city due to the its location along the main Yemeni Highland Pilgrim Route that connected Southern Arabia to Mecca (al-Thenayian (1997:244).

Historical records mention that Sa'dah was famous for the production of fruits, especially grapes (Gingrich and Heiss 1986:14) but orchards were found mainly in some nearby wadis where irrigation from spate and wells made fruit production possible.

There is no evidence to suggest that agriculture played a significant role in the basin itself. Insufficient precipitation and lack of surface water resources were constraints, factors which also explain why settlements were concentrated in or near wadi areas at the periphery of the basin.

However, the basin had large deposits of iron ore which have been mined over many centuries (Gingrich 1986:14; Meinhold, K.D. and Trurnit, P. 1981). Processing the iron ore was greatly facilitated by the abundance of acacia trees in the area. As a result many agricultural tools and weapons were manufactured for the wider region.

Historically, the Sa'dah basin is often mentioned for the raising of livestock, especially cows (Heiss 1987:64). Its large areas of grazing land and the abundance of trees and shrubs made this possible. Consequently, and as mentioned by the geographer Yakut (d.1229), Sa'dah became a commercial centre for the tanning of hides (Encyclopaedia of Islam 2 Vol. VIII:705). The preparation of leather was greatly helped by the fruit of the acacia trees (*qaraz*) which was used in the tanning process (Gingrich and Heiss 1986:15; Heiss 1987:65). According to al-Hamdani these trees spread as far as two days travel from the town and the whole area became consequently known as *bilad al-qaraz* (*qaraz* region, Heiss 1987:65).

A survey of available sources suggests that;

- Ecological conditions in the Sa'dah basin were major constraints to agricultural production but provided incentives for the rearing of livestock.
- Historically, pilgrimage, trade, mining of iron ore and tanning were major economic activities.

Until the early 1970s much of the Sa'dah basin was mainly grazing land. 'There was *mawt* (death) if one needed to make the journey on foot through the basin – only a few wells and hamlets on the way' remarked an older tribesman to the writer. And a foreign aid worker arriving in Sa'dah town in 1974 recalled that 'there was nothing but shrubs and wasteland as far as the eye could see.' What made agriculture possible under such circumstances?

The Sa'dah basin receives, on average, less than 200 mm of precipitation per year. This is insufficient for rain-fed agriculture. In spite of these climatic constraints the region produces a variety of cereals, fruits and vegetables. In the following sections the main types of water resources which local farmers utilise to ensure a crop will be outlined.

A number of reasons explain the intensive agricultural production which characterises the two major wadis that enter the basin from the south and south west, Wadi al-Abdin and Wadi Alaf which becomes Wadi Sahn. Their approaches from the surrounding mountains into the basin are gradual and their slopes are gentle. In contrast, flood waters from the two main western wadis, Wadi Sabr and Wadi Fallah, cascade down from greater altitudes and with considerable force. Wadi al-Abdin and Wadi Alaf are also closer to the markets of Sa'dah town.

Spate Irrigation

Spate irrigation (*sayl*) utilises flood waters that mainly occur during the two rainy seasons – March-May and July-August. This applies, in particular, to the wadi areas that enter the Sa'dah basin. Here, the production of fruit, especially grapes, has always been dominant (Heiss 1987:68). Both, Wadi Alaf and Wadi al-Abdin are mentioned by al-Hamdani for the production of grapes and other fruits and Wadi al-Abdin appears to have supplied the town of Sa'dah with fruits and vegetables (Heiss 1987:68). Historical evidence and oral tradition makes mention of a small dam in Wadi al-Abdin built in the 7th century. The dam supplied water to irrigate gardens just south of the Sa'dah town until it was destroyed in in the year 815/816 AD (Heiss 1987:66f). The existence of this dam might also explain why Sa'dah was described as a well watered city during that period (Encyclopædia of Islam 2 Vol. VIII:705).

In any given year wadi spates and their volumes are uncertain and there might not be sufficient water for all the parties involved. In this case Islamic and customary regulations stipulate that up-stream users are entitled to take whatever water they need first, down-stream users take what is left. This is known as *al-a'la fi-l-a'la* (Caponera 1973:16,18, 212). In many cases, such as in Wadi al-Abdin, this situation favours one tribal group with agricultural land up-stream over another which farms further down the wadi. Here, the people upstream are part of the Hamdan Bakil confederation while those down-stream belong to the Khawlan b. Amir.

For a number of reasons, however, this does not usually result in conflict over water resources. Firstly, communities accept the 'up-stream first' principle (*al-a'la fi-l-a'la*). Secondly, sub-surface water in these wadis flows at depths of 10–20 metres and hand-dug wells have traditionally provided supplementary irrigation. This has meant that even down-stream users have been able to obtain enough water to ensure a crop. Thirdly, land for agricultural use is limited to the narrow wadis and no expansion is possible. Therefore extensification of agriculture in these areas is not an option and water demand is constrained by a lack of land resources. And finally, grapes are produced on almost half of the land in the wadi zones. Vines are less water demanding than many other crops. The social and economic values of grapes will be explained in Chapter 5 but a few comments are needed here.

Sa'dah has a very long tradition of producing high quality grapes. Grapes are valued as both a cash crop and a subsistence crop. Fresh grapes as well as raisins from the Sa'dah region are sold at most Yemeni markets. Moreover, these products are also in demand in Saudi Arabia and returns in Saudi currency provide stable incomes. But importantly and as far as water requirement is concerned, vines can endure water stress and cope with drought over long periods: grape producers mention up to several years. No other fruit grown in the Sa'dah area, apart from qat, has this unique quality. And finally, grapes grown under a degree of water stress and without supplementary irrigation are perceived as better quality. Rainfed grapes are sweeter and fetch higher prices than those which are irrigated.

Until the early 1970s the wadi areas were the main centres of population. Here, all available agricultural lands have been utilised since time immemorial and expansion of agriculture is not an option. The mountain flanks are steep and established water rights do not allow for additional entitlements. Population increases have put enormous pressures on available land resources and forced people from the wadis to the basin to look for land.

Figure 3.9 (A) shows the result of Kopp's (1981:206) survey of these wadi zones in the Sa'dah area in the late 1970s. Field observations indicate that crop patterns have not changed significantly over the past 20 years. The area given to grape production has remained the same and still covers up to 50 percent of agricultural land in the wadi zones of the basin. Also important is alfalfa that is used as fodder for livestock. The adjacent mountain areas are often steep and barren, providing little extra grazing for sheep and goats. Moreover, wadi communities close to the town of Sa'dah grow alfalfa for the city dwellers, who all keep sheep and goats for milk and meat (Chapter 5).[3] After alfalfa comes wheat and sorghum, each occupying approximately ten percent of agricultural land. Pomegranates, fruits and vegetables, and qat make up the rest.

Water rights and runoff agriculture   Conditions are very different in the basin itself. During times when floods from wadis are strong and carry abundant water spates that enter the basin follow shallow depressions. Before the coming of tube wells in the mid-1970s patterns of agriculture and human settlement in the Sa'dah basin itself followed precisely along the direction and flow of these spates (Van der Gun 1985:20). However, most excess surface water rapidly infiltrates through the basin's alluvial layers and not many spates from the surrounding wadis reach the areas in the central basin. Historically, moving inwards from the periphery to the central parts of the basin, there is a progressive decline of agricultural activity and settlement density (Van der Gun 1985:16).

In spite of these environmental and geophysical constraints a limited amount of agricultural activity has been possible by applying rainwater harvesting techniques known as *sawaqi* (runoff agriculture). Eger's study (1984) in the Amran valley has shown that even in areas that receive as little as 150 mm of rain per annum harvests can be achieved by applying runoff techniques.[4]

In runoff agriculture rainwater is collected from unused agricultural land from where it is conveyed through shallow ditches or along low stone walls, and by using the natural forces of gravity, to adjacent cropped areas known as runon (Kopp 1981:110f). The runoff to runon ratio is fixed by the amount of precipitation that can be expected. In areas that receive only marginal amounts of precipitation, such as the Sa'dah basin, runoff areas can be many times the size of the receiving plot (Kopp 1981:110). For Yemen's Amran basin, for instance, Eger 1986:198;193) reports runoff-to-runon ratios of up to 20:1 on the alluvial deposits of the Amran valley.

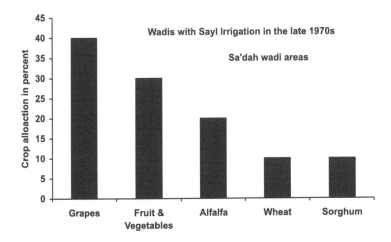

Source: Kopp 1981:206

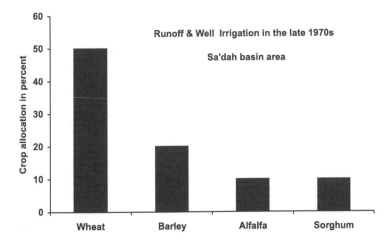

Source: Kopp 1981:207

Figure 3.9   **(A) Cropping patterns for wadi areas in the Sa'dah area in the late-1970s** (above)
(B) Cropping patterns for the Sa'dah basin area and wadis in the late 1970s (below)
Note that sorghum and wheat are usually grown on the same field, sorghum during summer and wheat in winter.

Not every individual plot has its own *sawaqi* but can serve a cluster of plots. *Sawaqi* have been built with great care and by benefiting from a detailed knowledge of the local terrain to ensure that excess runoff is diverted and/or channelled to secondary plots.

Without *sawaqi* irrigation a crop would be possible only where farmers could supplement irrigation from spates or wells. But with sufficient runoff two harvests can be achieved where otherwise hardly one could be expected (Kopp 1981:111). Moreover, runoff carries with it various beneficial organic substances, which are then deposited on the plots thus ensuring a level of soil fertility (Kopp 1981:111). *Sawaqi* are of such importance to the traditional agricultural system in the Sa'dah basin that they have become an integral part of the customary laws that regulate land rights (Caponera 1993:29). In this context Varisco (1996:242) notes that '[w]here land is virtually unusable without irrigation it is important that land and water be linked.'

Prior to the changes that occurred in the 1970s most land in the Sa'dah basin was *sabil* i.e. belonging to the tribe. If a part of it is cultivated it becomes *mulk* or private land. But as Kopp (1977:26) points out, in a strictly tribal sense, one can only obtain the right to utilise the land and only that right can be leased or sold.[5] A person who establishes *sawaqi* obtains for the entire runoff area the runoff rights. Consequently, land rights and water rights become one inseparable unit and can only be sold or traded together.

Until the mid 1970s land use in the Sa'dah basin was firmly governed by traditional runoff rights. Kopp (1977:26) has shown some of the implications that runoff rights can have. The area of a village is *sabil* i.e. belongs to the tribe, and any member has the right to build a house on it. However, he has to make sure that the structure does not infringe on someone else's water rights. Since virtually every square metre of land in the highlands (this applies to the Sa'dah basin) has runoff rights attached to it, the house has to be built in such a way as to allow the runoff from the grounds to continue to flow to the previously receiving plot. Runoff rights, Kopp argues, then also explain the absence of house cisterns for domestic water use in the region. The only exception are mosques where rain water is collected from roofs to be stored in cisterns and used for ritual ablutions. A new mosque can only be built on land where the benefactor holds the runoff rights. Consequently, the receiving plot will end up as a cemetery as it no longer receives sufficient runoff to secure a crop. This, according to Kopp (1977:26), explains the relative position and distribution of cemeteries in villages where runoff agriculture is dominant.

Prior to the mid-1970s perhaps as much as 90 percent of land in the Sa'dah basin was governed by established runoff rights. **Figure 3.10** explains the effective control exercised by down stream communities and actors over many of their up-stream neighbours. The following case study illustrates this point in more detail.

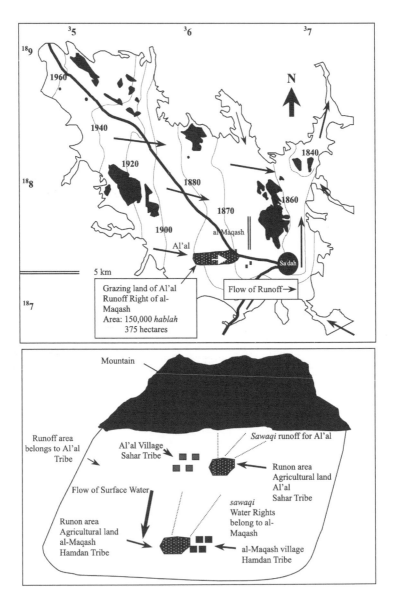

Figure 3.10   Customary runoff rights in the Sa'dah basin: the case of Al'al-al-Maqash

Water Rights and Tribal Control

The village of Al'al is part of the Bani Mu'adh, a subsection of the Sahar tribe, which in turn belongs to the larger confederation of Khawlan b. Amir. Al'al is located at the western edge of the basin, about 5 km off the mountains. Al'al owns a relatively large area of grazing land, about 150,000 *hablah* or 375 hectares that extend from the village eastwards, all the way to the main road that runs from Sa'dah town northwards through the basin. Near the road lies the settlement of al-Maqash. al-Maqash are not part of the Sahar, nor do they belong to the larger Khawlan b. Amir confederation. Al-Maqash are a sub-section of the eastern semi-nomadic Wa'ilah tribe which belongs to the confederation of Hamdan Bakil. Tribal belief is that some time in the distant past they left their ancestral territory in the arid east in order to seek refuge in the Sa'dah basin and the village of al-Maqash is said to be *hijrah* for Hamdan. As has been mentioned earlier, the basin slopes gently between 0.5–1 percent (Van der Gun 1985:6) from the west to the east, which means that the al-Maqash are down-stream of Al'al. The position of the al-Maqash in the centre of the basin requires a large runoff area for the small plots around their village. Therefore, al-Maqash have in the distant past established *sawaqi* that extend over the large grazing areas held by Al'al. The land belongs to Al'al but the runoff rights belong to al-Maqash.

Customary law stipulates that the right to runoff takes precedence over the right to the land (*al-sabb aqwa min al-salab*). Consequently, al-Maqash, although being an 'outsider' to the tribes of the basin, effectively controls the land use of communal land belonging to Al'al. The people of Al'al share the right to collect firewood and graze their flocks but are not permitted to do anything that might effect the volume and quantity of the runoff. Over the past two decades Al'al has repeatedly asked the permission of Al-Maqash to develop their communal land. Al-Maqash's refusal has led to numerous conflicts over the issue. Finally, an influential shaykh arbitrated between the two tribes and pushed for a postponement of the case for 20 years (*waqaf al-qadiyyah*).

Mediation by Religious Scholar in 1972 over Runoff Rights

Until 1972 traditional runoff rights constrained extensification of agriculture through the use of modern tube wells. Unlike many areas elsewhere in Yemen potential land for agricultural production was plentiful. However, for the reasons stated above Sa'dah's land resources could not be developed. At the same time, pressures on the existing and limited agricultural land resources grew due to population increase, technological changes and especially the growing awareness of Sa'dah's accessible groundwater resources. This situation resulted in many tribal conflicts throughout the area. Only through the arbitration of a religious scholar in 1972 was a settlement finally reached. His ruling became a mile-stone decision which opened the way for the expansion of agricultural development. He suggested that a community should render half of their grazing land to those owning the rights to the runoff from it. However, if those with runoff rights preferred the runoff no agricultural development could take place on the land. The scholar's arbitration was accepted unanimously by all the tribes in the area.

Subsequently, many tribal communities privatised their common land. Each household received land according to the number of males, children or adults. Households who previously had as little as one or two hectares had their land holdings increased by many fold. There was more land than individual families could develop. And as we shall see in the next chapter, for a number of reasons, this situation triggered the sale of land to members of other tribes and those seeking economic fortunes in the Sa'dah basin in response to the unprecedented cross-border trade and the boom of tribal markets witnessed during the late 1970s and early 1980s.

However, the 1972 arbitration also exacerbated old tribal disputes over land claims, especially where boundaries between tribes and within communities had never been very clear-cut. Disputes over land intensified as land values soared. If large areas of the land have been left uncultivated until the present time it is to avoid tribal conflict over territorial boundaries. Paradoxically, these are the areas where the Sa'dah aquifer has escaped exploitation. As the local communities have become increasingly aware of the rapidly falling groundwater table, and as costs to run and maintain pumps and equipment have become unaffordable, traditional rights to runoff are now preferred to the gains of additional land. Returning to the case study, al-Maqash are unlikely to concede their runoff rights even though they would gain a huge amount of territory. The future of groundwater irrigated agriculture is too uncertain and the present arrangement leaves them in control over the resources of their tribal adversaries. They reckon that this bargaining power might well become very useful in the future.

It has become evident that most of the land in the Sa'dah basin was used as runoff areas. In spite of these *sawaqi* runoff systems prolonged periods of drought presented real risks to agricultural production.[6] In contrast to the wadi areas described above; cereals were the main crop in the basin. The farmers of the region raise many different varieties of sorghum, wheat and barley, which can cope with the variations of rainfall. **Figure 3.9** (B), presented a few pages earlier, shows agricultural patterns of the 1970s in the basin. These figures reveal some interesting comparisons to those on the wadi areas seen earlier. In the basin, cereals are grown on about 80 percent of agricultural land. The lower figure of ten percent for alfalfa indicates that natural grazing was sufficient to provide fodder for livestock. It is noticed too, that wheat took precedence over sorghum during the 1970s, a trend which was reversed during the following two decades. Cheap imported wheat flooded tribal markets during the 1980s and removed incentives to grow local varieties (Morris 1986:138f, 172; Weir 1987:280–281), while sorghum now provides essential fodder for livestock as much of the natural grazing land has been lost to groundwater irrigated farms. Kopp (1981:207) also observes a trend from cereals to fruit production, as pump irrigation reduced the risks of plant stress, and in response to the availability of cheap imported wheat. At the time, many farmers responded to market signals and started to produce cash crops.

Concluding this section   In the past, perceptions of the value of water in the Sa'dah basin have been shaped by water scarcity, tribal notions of political and economic autonomy, and by customary and Islamic law. In response to ecological constraints and exacerbated by prolonged periods of drought, the tribal population of the Sa'dah basin in the past developed efficient rainwater harvesting systems to grow cereals, alfalfa and produce quality fruit, especially grapes and pomegranates. However, it is doubtful whether the area's farming communities ever were food self-sufficient for any length of time. Until 1933/34 when Yemen had to concede its two northern provinces Najran and Asir, Sa'dah was a centre for trade, commerce, the tannery industry and the mining of iron ore. Migration, trade and livestock production have been common responses to supplement incomes and diversify livelihoods.

Politics and Actors

Republicans and Royalists

Sa'dah has been the religious, and often political centre of many Zaydi Imams (Gingrich and Heiss 1986:15). From the end of the 9th century AD when Yahya b. al-Husayn arrived in Sa'dah and mediated tribal disputes (Heiss 1987:74; Dresch 1993:167) the northern tribes have been associated with Zaydism (Dresch 1993:11). The two main confederations of Hashid and Bakil are even called the 'wings of the imamate'.

Tribal support for the Imamate has usually been conditional upon non-state interference in the affairs of the tribes and in exchange for autonomy and independence. During the civil war some tribes of the Sa'dah region supported the Republican cause, while others fought on the side of the Imam. Bribes and gifts, especially weapons, often determined which side people were fighting for (O'Ballance 1971:81).

Initially, there was little support by the northern tribes and their shaykhs for the Royalist cause. Their enthusiasm only changed after the Imam had secured large sums of cash from Saudi Arabia with which to bribe the tribes (O'Ballance 1971:81). The Republic, too, bought the support of many northern tribes with food, arms and money (O'Ballance 1971:180). In some cases this meant that tribes were rewarded as long as they stayed neutral and abstained from supporting the Royalists. In other cases '[N]othing was given or paid out by either the UAR or the Republican Government, unless for some service rendered. For example, the Hashid federation received money to 'guard' communications, especially between Sana and Sa'dah, and any incident caused payment temporarily to cease. This policy was quite successful and influenced a growing number of tribes to come over to the Republican side, to be rewarded and left alone instead of continually harried from the air' (O'Ballance 1971:180).

Sa'dah fell to the Republicans on the 3rd of September 1969 only weeks after the Royalist, prince Abdullah Hasan was murdered on his way to Friday prayers. The end of the revolution also marked the beginning of a struggle for influence and power in the Sa'dah area. According to a leading tribal shaykh and member of parliament, this struggle was a predominantly tribal one, whether Sa'dah would be *Hamdaniyya* or *Sahariyya*. i.e. whether the tribes of Hamdan Bakil from the east or the tribes of Sahar from the Khawlan b. Amir confederation would take control of the town and its immediate surroundings. In the following discussion some of the main actors in the struggle for power and influence in the Sa'dah basin will be considered.

Shaykhs and their Quest for Power

Table 3.1 shows the different tribal leaders, their social, political and economic position, and importantly their control over natural resources, which is summarised.

Table 3.1   Influential actors of the Sa'dah region, their social, economic and political position

Actor	*Tribal*	*Politics*	*Economy*
Dirdah b. Ja'far	Shaykh (Sahar)	Royalist	Political subsidies, Farm
Husayn al-Surabi	Shaykh Bani Mu'adh (Sahar)	Republican	Political subsidies Large farms Hotels (Sa'dah, Sana'a)
Husayn Fa'id Mujalli (died 1997) Uthman Mujalli (son of Husayn)	Shaykh Rahban, al-Abdin, Ghuraz, Farwa Shaykh	Republican Political Security Member of House of Representative (MP)	Political subsidies Real Estate Hotel
Faysal Abdullah Mana	Shaykh al-Talh	Member of House of Representative (MP) (al-Talh area)	Trading
Ahmad Hamdan Abu Mus'if	Shaykh Wa'ilah	Member of House of Representative (Kitaf area)	Political subsidies Trading Farms

Actor	Tribal	Politics	Economy
Abdullah Hamis al-Awjari	Shaykh Wa'ilah (Hamdan Bakil)	Royalist	Political subsidies Large farm in Sa'dah basin
Salih Salih Hindi Dughsan	Shaykh Al Ammar (Bakil)	Member of House of Representative MP (GPC)	Political subsidies Farm in Sa'dah basin
Qa'id Shawit	Shaykh Bani Uwayr	Republican Formerly member of House of Representative (Islah), now Member of Consultative Assembly	Political subsidies Military Trading & Contracting
Nasir b. Qirshah	Shaykh Walad Mas'ud (Sahar)		Political subsidies Farm
Traders	Close links with Shaykhs	Close to Gov. sources	Cross-border activities International trade

Note: All these shaykhs receive regular moneys from both, Yemeni and Saudi political sources. MP stands for elected member of the House of Representatives (*majlis al-nuwwab*) as opposed to a member of the Consultative Assembly (*majlis al-shu'ara*).

As stated above, most of the Sa'dah basin is part of the territory of the Sahar tribe. Their overall shaykh is Dirdah b. Ja'far. He supported the Royalists during the Revolution and subsequently lost bargaining power with the Republic. However, he continues to receive regular salaries from Saudi Arabia for his earlier loyalty to the Royalists. Visitors to his humble home in Harf Ja'far, at the western edge of the Sa'dah basin (**Figure 3.11**), are usually shown a little piece of paper signed by the President who, during a recent visit to Sa'dah, made a small 'gift' of money to shaykh Dirdah. Older tribal people still pay him respect and on special religious occasions shaykh Dirdah is saluted with guns fired into the air.

Real power has long moved to Husayn al-Surabi whose village is just a few kilometres south of Harf Ja'far. His large land holdings are symbolic of his wider influence and his rise to prominence needs to be considered in more detail.

Shaykh Husayn al-Surabi of Bani Mu'adh   Husayn al-Surabi is the present shaykh of the Bani Mu'adh tribe, a position he inherited from his uncle Faysal al-Surabi when the latter died from a liver disease in 1983. A tribal incident in 1986 (explained below) became instrumental in strengthening his leadership among the

Figure 3.11   (A) In conversation with shaykh Dirdah bin Ja'far
Traditionally the shaykh over all the Sahar tribal sections bin Ja'far
has gradually lost political power since the 1962 revolution.

Figure 3.11   (B) The old home of Bin Ja'far in Wadi Harf Ja'far
His house (left on the photo) is now unoccupied. Bin Ja'far and his
family have moved to a new location closer to the main Sa'dah-Saudi
road where water is more plentiful to irrigate their new farm.

Bani Mu'adh. Moreover, his role and courage throughout the subsequent settlement, which lasted over two years, established his reputation and gave him political clout far beyond his tribal boundaries.

In 1986, two men from Hashid killed two men from Bani Mu'adh in Sa'dah. The offenders were apprehended and put in prison in Sa'dah. In an attempt to avoid a blood feud, the relevant Hashid tribal section offered to meet the Bani Mu'adh on their home turf in order to sacrifice and pay retribution. However, the Bani Mu'adh rejected the blood money. They judged the case as intentional killing (*amdan*). Consistent with tribal custom and in order to avoid 'shame' the Bani Mu'adh requested that the perpetrators be killed.

When Abdullah b. Husayn al-Ahmar,[7] the present political leader of the Islah party and the paramount shaykh of the Hashid confederation of tribes intervened on behalf of the offenders, the Bani Mu'adh took matters into their own hands in order to avoid an outcome that would be unacceptable to them. Relatives of the two victims manoeuvred their way into the Sa'dah prison and the killers were put to death. The case exposed the limits of government control even within the walls of its own institutions and developed into a political issue with the Hashid leadership accusing the government of weakness and negligence. When shaykh Surabi of Bani Mu'adh was summoned to travel to the capital to explain the situation, he bluntly refused to go, saying that 'we will not answer the government, if Hashid wants to fight let them come and we will be ready' *(la nujawib al-dawlah wa lakin in kan Hashid tishty al-harb fa ahlan wa sahlan).*

In a show of power the army surrounded the prison with tanks in order to capture those bound on revenge from the Bani Mu'adh who were trapped inside. They had vowed to fight until the last bullet but shaykh Surabi persuaded the men to give themselves up. Initially they were taken to the capital where they were later given the death penalty in a San'a court. However, through the bold intervention of shaykh Surabi who went to San'a to confront the President over the issue, the two men were eventually released and flown back to Sa'dah in order to avoid capture on the overland route through Hashid territory.

Essentially, the incident can be seen as a confrontation of power between Yemen's most powerful tribal shaykh, Abdullah b. Husayn al-Ahmar, representing central state power, and shaykh Surabi representing local Sa'dah tribal power. At a time when the government could not afford to alienate the tribal groups in the Sa'dah basin and was trying to consolidate its grip over the Sa'dah region it is possible to speculate that a number of political considerations helped turn the balance in Surabi's favour. Moreover, Surabi's argument was perfectly in line with tribal logic; 'those who died from Hashid are equal in number to those who died from Bani Mu'adh' *(alladhi matu min Hashid mitl alladhina matu min Bani Mu'adh).* A further argument put forward by Surabi stressed the notion of tribal equality criticising the political leadership (the majority of which are from Hashid tribe). – 'the blood of those killers in prison is not worth more than the blood of our own two men' *(ma dam al-nas fi-l-sijn aghla min haqqana).*

Shaykh Surabi's position with the president and with the Sa'dah tribes was enormously enhanced as a result of this incident. Apparently, sections of his tribe rewarded his courage against the 'power' of shaykh Abdullah (b. Husayn al-Ahmar) by presenting him with land while it is said that the President acknowledged Surabi's sphere of influence by sending out a drilling rig to reward him with a number of tube wells.

Unlike a number of the other shaykhs in the basin, Surabi has kept clear of party politics. This, perhaps, explains his role as a mediator between other local shaykhs. Honour comes from being associated with great shaykhs as Dresch (1993:101) makes clear, and one important indicator of a shaykh's position is in the number, frequency and kind of guests he hosts. Among the northern tribes there is perhaps no one like the almost legendary Mujahid Abu Shawarib (Dresch 1993:101, 362) who commands the respect of both Bakil and Hashid tribes. The fact that Mujahid stays with Surabi during visits to the Sa'dah basin only favours Surabi's position.

Following the 1984 import ban for fruits and vegetables, Surabi was one of the first to start planting citrus orchards. More than any one else in the area, his name and reputation are associated with large scale irrigated farming, and over the years Surabi has used his power and position to increase his land holdings in the Sa'dah basin. However, his quest to own and control increasing areas of the basin's land and water resources has not always been realised. On a number of occasions individual stakeholders and communities from his own tribe have used traditional tribal values and local knowledge to stop him from expansion.

Shaykh Mujalli of Rahban, al-Abdin, Ghuraz and Wadi Farwa   Until his sudden and untimely death from heart failure during the spring of 1997, Husayn Fa'id Mujalli was shaykh of an area just south and south-west of Sa'dah town. His son Uthman, has since taken over the leadership from his father. The territory under the influence of the Mujallis is part of Sahar. It includes some of the most fertile wadis, famed for their grapes and pomegranates, Rahban, Wadi al-Abdin, Ghuraz and Wadi Farwa.

Husayn's father Fa'id supported the Revolution and rewarded himself accordingly. At the end of the civil war he appropriated large *waqf* holdings which had been associated with certain *sayyid* families for centuries.[8] It is no secret that he 'acquired' (*iktasab*),[9] or so many people think, much of the *waqf* land (religious endowment) that stretched from the town's southern gate (Bab al-Yemen) westwards to what is now the new town with its Saudi financed hospital, its governmental offices, shops and blocks of flats.

The late Husayn Fa'id Mujalli has not focused his efforts on agricultural activities but has instead used his power and influence to manoeuvre himself and his family into positions of political and economic power. The combined value of his land and property holdings (formerly *waqf*) along the main road through the new section of Sa'dah town are enormous. His real-estate includes a large hotel, government buildings and many shops and flats.

His son Uthman is well positioned to carry on his father's torch. In the 1997 parliamentary elections Uthman defeated his rival, shaykh Qa'id Shawit (Islah Party) who was consequently relegated to become a Member of the Consultative Assembly.[10] He is close friends with one of the President's sons, who arrived by helicopter to attend the funeral of Uthman's father.

Shaykh Abdullah Hamis al-Awjari of Wa'ilah (Hamdan) Shaykh Abdullah Hamis al-Awjari is one of the prominent shaykhs of the Wa'ilah tribe (Hamdan Bakil) whose territory extends from the basin's eastern periphery along the Yemeni-Saudi border up to Najran. The home of shaykh Abdullah is in Wadi Nishur, north-east of the basin but since the mid-1980s he has managed to appropriate considerable land holdings just two km north of Sa'dah town where he now irrigates 25,000 *hablah* (62.5 hectares) of citrus. The details of how he succeeded to establish control over land, claimed and contested by rival shaykhs from Sahar tribes are better left undisclosed. But a survey of his farm quickly reveals some of the risks associated with the quest for power in this region. Next to one of the wells lie the remains of a Toyota Landcruiser (**Figure 3.12**). Only a few years ago a land mine, meant for the shaykh, mistakenly blew up his son instead as he approached the farm. It then comes

Figure 3.12 (A) Wreck of a Toyota land cruiser
The car had run over a mine near the shaykh's farm. It was an attempt to kill the shaykh but killed his son who was driving the vehicle at the time.

Figure 3.12   (B) Shaykh Awjari's citrus orchard

as no surprise to see a man like shaykh Abdullah constantly surrounded by a number of body guards.

Unlike the Sahar shaykhs (who gained influence and power through supporting the Republic) Abdullah's father was instrumental in organising Royalist opposition in order to stop the advance of the Egyptian army in the direction of Saudi Arabia. According to tribal wisdom, this explains why his 'regular salary', which he receives from sources across the border, is three times that of other shaykhs in the area.[11]

When his citrus trees are hit by a disease the shaykh sends samples north to Saudi Arabia and not south to San'a for analysis. Fertiliser and pesticides too are brought across the border, a further indication of his political alliances. However, shaykh Abdullah's large citrus farm is only symbolic of his influence. The real power of the Wa'ilah shaykhs is derived from the fact that they control the main routes that pass from the Sa'dah basin to Saudi Arabia (Dresch, 1993:309, 348, 381; Gingrich 1993:263).[12] Tribes and traders alike depend on the Wa'ilah for the safe passage of their goods. This was especially critical during the height of the cross-border trade in the late 1970s and early 1980s when just about everything was smuggled across from Saudi Arabia, from petrol to TV sets and irrigation equipment. Tribes as powerful as Hashid, who were in the business of moving and distributing the goods, relied on Wa'ilah, a sub-tribe of Hashid's rivals, the Bakil, for guaranties and securities.

Shaykh Abdullah's farm is just north of the road that leads east through his home turf of Wadi Nishur to Kitaf and on to the border crossing point at Najran. The road has recently been asphalted making it possible to reach Kitaf in just 45 minutes. From there 'all roads lead to Rome' (Saudi Arabia). The sweet scent from his citrus blossoms during February and March fills the area and can be smelt from a far distance. To many, his gardens evoke visions of the gardens of paradise.

During the mornings shaykh Abdullah can usually be found on his farm where he divides his time between inspecting the trees and attending to tribal matters. Meeting him early one morning in January 1999 concerning the kidnapping of a colleague and friend by a Bakil tribe from the east, it became clear that he carries considerable tribal-political clout. At the same time his influence at the national political level is limited because of his past and his questionable loyalty to the state. 'If the kidnapping were a tribal affair we (meaning the Wa'ilah tribes) would negotiate the release of your friends' he said. 'However, as the case involves the government our hands are tied and we just have to wait and hope for the best outcome'.

Shaykh Salih Salih Hindi of Al Ammar. Salih Salih Hindi is shaykh of the Al Ammar, a Bakil tribe whose territory lies just south of the Sa'dah basin. The main road from Sa'dah town to the capital San'a passes through Al Ammar, a factor which provides them with some control over the comings and goings to the Sa'dah basin. Moreover, the government's main customs duty point is within Al Ammar's territorial boundaries. If those transporting goods get delayed at the custom point, or if they feel that the custom's officers demand unfair duty fees, they will come to shaykh Salih and ask him to solve the matter.

Al Ammar's territory also extends to Wadi Madhab which runs south east and drains into the Jawf graben. Wadi Madhab provides a natural route all the way through Bakil territory to the Jawf and onwards to the Empty Quarter and the regions of the south. Salih's father who, to the surprise of most in the area, died of natural causes in the autumn of 1995 and not as a result of tribal revenge, was involved with the socialist movement in the former South Yemen. Consequently, Al Ammar's relationship with the government can only be described as one of mistrust. Nevertheless, Salih, like his rival Uthman from Rahban, is an elected member of the House of Representatives (*majlis al-nuwwab*), for the People's Congress Party. Like many of the other 'big' shaykhs from the Sa'dah basin Salih divides his time between the capital and his tribal home at Al Ammar.

Intra-tribal as well as inter-tribal tribal rivalries and feuds demand that Salih only travels in the company of his armed bodyguards. An attempt in 1989 to kill Salih's father inside the perimeter of Sa'dah's government buildings failed, but a number of men from both sides were killed and many were wounded. Following the incident, the government literally besieged the shaykh's house at Al Ammar with tanks and artillery in order to prevent the outbreak of a major tribal conflict.

Most of Al Ammar's terrain is rocky, barren and arid. Agriculture is confined to a few wadis such as Wadi Sharamat. Population increase puts pressures on available land resources. Outside of Wadi Sharamat there are hamlets in some of the depressions. Here, some grapes and qat are produced by utilising runoff. Wells have to be drilled to considerable depths, about 350 metres through base rock, to abstract sufficient volumes of groundwater, investments justified only by qat cultivation. The only suitable area for agricultural expansion appears to be Wadi Madhab which collects rain and flood water from a large area and where groundwater can be found at only 10–20 metres depth.

Wadi Madhab is communal property of Al Ammar but the land has not been divided according to the number of male members. Also, Wadi Madhab is at a lower altitude, hot, humid and malaria infested and therefore Ammaris avoid living too long in the wadi.

The Sa'dah basin is a much more suitable choice for migration and many Ammaris have moved there. Shaykh Salih, too, has a large citrus farm (7.5 hectares) within the territory of the Bani Mu'adh. He hardly visits it, but what counts is that he erected a visible landmark to his presence. The location of his farm and its tribal-political context are significant. Salih's farm is in the south-western corner of the Sa'dah basin near the village of Mahdah. Mahdah, like al-Maqash described earlier, is a settlement of the Hamdan Bakil among the Sahar tribe of the Khawlan b. Amir confederation. Like al-Maqash, Mahdah claims the runoff rights of a considerable area. This has led to numerous conflicts with sections of Sahar tribes over the past two decades. Surrounded by the Sahar Mahdah felt vulnerable. Its members were especially concerned about 'losing' their claimed rights to land and water resources. In a move to strengthen their position vis-à-vis the Sahar, Mahdah transferred or sold some of their land to fellow Bakil tribes, including, especially, people with perceived tribal-political clout and influence, such as shaykh Salih of Al Ammar.

Qa'id Shawit of Bani Uwayr   Towering over Al Ammar on a high and separate mountain plateau to the west live the Bani Uwayr who are part of the Sahar tribes. Their plateau is accessible only by two steep mountain tracks. Rainfall is sporadic and unreliable and agricultural activity is confined to wadis and depressions where runoff from the barren hills allows for the production of grapes and cereals. A visit to Bani Uwayr quickly reveals that the livelihoods of most are not in agriculture. Most four-wheel drives on the plateau show military number plates, an indication that many of their able men are on the payroll of the army. This is not surprising since their shaykh, Qa'id Shawit, established close connections with the military through one of his sons and a nephew holding higher positions in the army.[13] By contrast, Bani Uwayr's former shaykh wavered too long during the Revolution and then leaned (*yamil*) toward the Royalists. As a result, he gradually lost influence at the political level and, as a consequence, at the economic level.

Many communities living on his side of the plateau still consult him on matters of tribal concern while they approach (officer) Shawit about issues concerning wider national and political affairs. Shawit's political connections also suggest reasons why his eastern side of the plateau is better developed in terms of domestic water supplies, telecommunications and infrastructure.

Qa'id Shawit's standing and reputation also stems from the fact that he acts as mediator for the Hamdan Bakil tribes. The hard lessons of how to solve tribal conflicts have, perhaps, been learned at home. Bani Uwayr has been involved in a number of drawn-out tribal feuds, resulting, in one case, in the death of 20 people including two women.[14] Consequently, tribal sections involved in the dispute have had to stay put on top of the barren plateau for prolonged periods (one intra-tribal conflict between two section of the Bani Uwayr lasted ten years) to avoid being killed in revenge.

It is illuminating that a number of Sa'dah's most successful entrepreneurs and businessmen come from the barren plateau of the Bani Uwayr. In partnership with shaykh Qa'id Shawit these trading and construction companies have built hospitals, government facilities, schools and water supply systems.

Qa'id Shawit was MP at Yemen's House of Representatives (*majlis al-nuwwab*) representing the religious Islah party until the recent elections in 1997 when he lost his seat to shaykh Uthman Mujalli (Peoples Congress Party) for reasons which are worth summarising.

Radman, the son of Bani Uwayr's former shaykh stood against Qa'id Shawit in 1997, if only in an attempt to keep the number of potential votes for Shawit low. As a result Shawit lost his seat in the House of Representatives (*majlis al-nuwwab*) to Uthman Mujjalli, the son of the late shaykh Husayn. Locally, it is believed that shaykh Husayn Fa'id Mujalli cultivated a friendship with Shawit's opponent so that he could stand in the election changing the balance of votes in favour of his son Uthman. Exploiting the long-standing rivalries of Bani Uwayr's two shaykhs paid political dividends for the Mujallis.

Times of relative peace and security find Qa'id Shawit in his home just outside of the old town of Sa'dah and close to Mujalli's residence. Here Qa'id Shawit owns a farm, receives guests and stays close to political and economic developments. Also, here he is surrounded by many of his own tribal people who have left Bani Uwayr's meagre homeland to settle around Sa'dah in search of livelihoods, land and water resources.

One small incident illustrates the rivalries between these shaykhs in their quest for political power and economic control over Sa'dah's natural resources. The author had planned to pay a visit to Qa'id Shawit. As his intention became known a local man from the area readily offered to drive him there. On the way the man stopped at the Government's Court House (*mahkamah*) where these shaykhs spend a good deal of their time 'sorting out' land issues and title deeds. A few minutes later the late shaykh Husayn Fa'id Mujalli came out to welcome the 'foreign specialist' (*khabir*). As a result, lunch and afternoon qat chew were spent (refusing the hospitality of a shaykh is not a wise option), not as intended, in the company of Qa'id Shawit but shaykh Husayn Fa'id Mujalli. One of his agents had deliberately hijacked the visit, no doubt, in an attempt to steer potential 'foreign' benefits to their patch. Not surprisingly, shaykh Mujalli took the foreign 'khabir' to where the legendary dam in Wadi al-Abdin is believed to have been (Heiss 1987:66). His proposal and request was for a new dam to be built, this time near his new home, which would be the most suitable location, or so he figured.

al-Mahadhir   Bordering Al Ammar to the north and Bani Uwayr to the north-west is the area of al-Mahadhir, a separate basin but much smaller than Sa'dah basin. Al-Mahadhir receives much of its runoff from the Bani Uwayr plateau about 300 metres above al-Mahadhir. Runoff rights have, so far, prevented privatisation of the large communal grazing areas. And there have not been the same economic and political dynamics as in the Sa'dah basin to justify the expansion of irrigated agriculture. As in the case of their neighbours, the Bani Uwayr, Mahadhir is home to a number of entrepreneurs and businessmen, a subject we will return to when we consider how communities adjust to water scarcity.

The new road from the capital San'a to Sa'dah, completed in 1979, according to Meyer (1986:265) passes right through al-Mahadhir linking the area with the Sa'dah basin in only 15 minutes by car. Nevertheless there are incentives to live in Sa'dah and al-Mahadhir's shaykh Abu Ubayd Bushayt has his house near the Saudi financed Salam Hospital at the western end of Sa'dah's new town. Many of his fellow Mahadharis have followed him there. Shaykh Bushayt's economic and political position does not appear to be as prominent as some of the other shaykhs, indicated by the fact that the area still lacks many basic services, such as water and electricity supply.

As mentioned earlier, al-Mahadhir has large grazing areas and those that have moved into Sa'dah town have brought their livestock with them. It is not unusual to find single households keeping between 20 and 30 head of livestock. In the absence of the traditional grazing areas fodder has to be bought from irrigated farms close by.

Traders

Besides Sa'dah's tribal shaykhs, some of who have gained considerable influence at both regional and national level, the basin is home to a number of rich and influential trading families. In many cases these actors have formed partnerships with the relevant shaykhs to combine political muscle with economic opportunities.

Migration into Saudi Arabia and the subsequent remittances sent back from the Kingdom created an unprecedented flow of goods and services, especially in Sa'dah's border areas. For economic and political reasons it became very lucrative to live near the booming tribal markets of the Sa'dah basin where many of these 'smuggled' goods were traded and from where they were transported to other parts of Yemen.

Today's prominent trading families made fortunes during the period from the mid-1970s till the mid-1980s. In several cases the origins of their economic power go back to the period of the civil war when trade in all kinds of weapons meant huge profits to those with the right links. Sa'dah's tribal setting as well as its geographical location provide explanations. Since 1934 several tribes of the

Khawlan b. Amir have lived on either side of the present border. A similar situation exists in the east among the tribes of Hamdan Bakil where tribal territories and loyalties extend across political boundaries. Local traders and shaykhs have been able to capitalise on the competing and often conflicting interests of both, the Saudi and the Yemeni governments.

Traders and Markets

To most Yemenis, and certainly tribal folk the word al-Talh stands for the largest tribal market in the country. Al-Talh conjures up notions of tax-free goods (smuggled goods), a mecca of the weapons trade,[15] tribal autonomy and the absence of government regulations – simply *the* place where tribes do business on their own terms. Al-Talh is located in the centre of the Sa'dah basin 15 km north of Sa'dah town and one kilometre east from the main road that runs north through the basin. It is here, around the market of al-Talh, where some of the most influential traders have established themselves.

Much of their initial economic fortunes are said to be related to the weapons trade. Not surprisingly, Yemen has apparently now made its entrée into the Guinness Book of Records with an estimated 50 million rifles being held by a total population of 16 million. *Allahu a'lam* – Allah alone knows the true figure.

While a number of these entrepreneurs have established international trading houses and businesses abroad, they have nevertheless remained devoted to the Sa'dah region where they spend much of their time. From their homes in the rural area around al-Talh and Sa'dah town, modern means of communication allow them to stay in close contact with both their regional and international offices. Their large homes, some of which resemble fortresses surrounded by high walls, are symbolic of their economic power. At the same time, their sitting rooms (*mafraj*) are open to anyone who wants to join the afternoon qat chew and discuss the issues and concerns of the day.[16]

Traders have learned to convert economic muscle into political power by living and operating in Sa'dah's politicised environment. Trading no longer appears to be dishonouring in tribal eyes. On the contrary, some influential traders are involved in tribal mediation and conflict resolution like shaykhs, often over land and water resources. A kidnapping in January 1999 illustrated their relative influence and power at both the local and national level. In an attempt to accelerate the release of a number of foreign hostages held by a tribal group in the east, a trader was visited by a resident western expert in the company of the author. After explaining the incident, the trader left the 80 men who had gathered that afternoon at his residence, and moved to one of his 'private' phones. Within minutes one of his relatives in the capital was instructed to meet immediately with the minister in charge of affairs in order to determine whether the stalled negotiations between the kidnappers and the government could be resumed by him, his 'friends' and his tribe.

The political reach of the Sa'dah traders is immense. Like some of the area's shaykhs they have major social, economic and political influence in the basin. At the same time they can also make things happen in San'a. Moreover, they operate effectively in trade relations.

The traders of the Sa'dah basin maintain social, economic and political ties with groups and individuals along both sides of the Saudi-Yemeni border. One family has sought links through marriage with a daughter of one of the main shaykhs of Barat, the home of the eastern tribes of Dhu Muhammad and Dhu Husayn which have gained a reputation for their unruly conduct vis-à-vis the San'a government (see also Dresch 1993:28).

The recent reception of the ruler of Najran, which is one of the main border crossing point east of Sa'dah, by one of Sa'dah's proverbial trading families also symbolises these traditional links. A wide staircase was built especially on the hillside of one of the trader's irrigated farms to give the Saudi emir a vision of Sa'dah's green paradise (**Figure 3.13**). Traders and entrepreneurs from the Sa'dah basin also frequently appear in the company of the President: his economically motivated visits to European countries, including the UK are just recent examples.

Figure 3.13   Farm with staircase
One of five or more large farms owned by one of Sa'dah's most influential trading families with the staircase leading up to the white building. It was constructed to receive the emir of Najran during his visit to the area in 1997.

As far as natural resources are concerned, from the mid-1970s on, Sa'dah's traders have been the principal importers of pump and irrigation equipment. Like most consumer items, this equipment was 'smuggled' across the Saudi-Yemeni border. The same trader families own the drilling rigs. In short, expansion and development of irrigated agriculture has directly benefited Sa'dah's trading families and increased their financial returns.

All trading families now own comparatively large farms where they grow mainly citrus and other fruits. One trading family shares 8 irrigated farms between their 4 sons. Moreover, traders were the first to gain experience with modern irrigation technology, no doubt in an attempt to assess marketing potential.

Conclusions and Summary

The aim of this chapter has been to introduce relevant aspects of the Sa'dah basin's environment, society and economy. It has been shown that scarcity of rainfall and lack of surface water resources until the mid-1970s, limited and defined agricultural activity. In the past, scarcity of water resources shaped and determined not only population density and agricultural settlement patterns, but economic activity as well. Intensive agriculture was mainly confined to some wadi areas, while rearing livestock was prominent in the basin itself. However, the evidence suggests that throughout the Sa'dah basin's history and even today, sustainable livelihoods are mainly found in trade, provision of services and in small-scale manufacturing.

In addition this chapter has explained how Islamic and tribal-customary laws pertaining to the allocation and management of the Sa'dah basin's scarce surface water defined, up to the mid-1970s, the relative power of one actor over the environment of another. It was shown that the power of down-stream users with established rights to runoff water effectively controlled and limited any change in land use on the runoff area, which in most cases, belonged to other actors upstream. Resulting tensions were exacerbated by the fact that the Sa'dah basin is shared between communities identifying with a number of tribal groups.

Meanwhile, Sadah's Wajid aquifer offered potential for groundwater abstraction. Groundwater abstraction and privately owned tube-wells offered one possible way to resist control by other actors. Furthermore, privately owned wells also meant an end to drawn-out and costly tribal conflicts over scarce surface water resources. Moreover, groundwater development promised secure water supplies leading to greater autonomy and food self-sufficiency, ideals which are highly valued in Sa'dah's tribal society (Adra 1985:275, 280). However, extensification of ground-water irrigated agriculture was, with a few exceptions only, impossible since a) most land was communally owned and managed and b) extensification was blocked by the power and control exercised by down-stream owners of runoff rights.

This chapter has shown that a solution to the impasse was found in 1972 by transferring water using rights, which consequently altered the control and balance of power with regard to land and water resources in the Sa'dah basin. What is significant in the context of this study are the consequences of the 'solution', i.e. expansion of irrigated agriculture was driven, to some extent, by the emergence of another actor – the state. In this context the role of some of the main influential actors – shaykhs and traders – were identified. Their quest for control over the resources of others (resource capture) has been hinted at but more details will emerge in subsequent chapters which will also analyse the ability of grass-roots actors to resist them.

As the next chapter will show, privatisation of communal lands and the consequent extensification of groundwater-irrigated agriculture can be understood as an indirect result of Sa'dah's politicised environment. Notions of power and interests, control and resistance explain expansion and development of groundwater irrigated agriculture.

Notes

[1]   Source, internet: www.al-bab.com/yemen/soc/qat

[2]   The purpose of *juwarah* tribal customary regulation is to prevent the sale of tribal land to people from outside the tribe. Tribal autonomy and food self-sufficiency can only be maintained by controlling the essential means of production, land and water, from outside influence. A tribal member who wants to sell a piece of land must first offer it to his own family, his neighbours and his tribe. Only after no buyers come forward can the land be offered to some one from outside the tribe (Kopp 1981:136).

[3]   Alfalfa is mentioned as one of the principal crops grown in the Sa'dah basin along with grapes, other fruit, vegetables and qat (Van der Gun 1995:16).

[4]   Water requirements for a wheat crop are about 600 mm depth (Allan, personal communication).

[5]   A similar notion is found in The Holy Bible (Book of Leviticus 25:23) 'The land must not be sold permanently, because the land is mine and you are but aliens and my tenants.'

[6]   Grohmann (1933:33f) mentions that drought in Yemen is a constant threat to agricultural production. A long period of drought in Arhab and Hashid around the end of last century (1884) let to the destruction of the vineyards which had existed there in the 1860s. He also reports that Wadi Bayhan received no rains for four years, between 1894 – 98 (Grohmann 1933:35). Another prolonged period of drought, which afflicted many regions of Yemen, began in the 1950s and lasted right through the civil war until the early 1970s (Stookey 1978:275). Parts of Barat, to the east of Sa'dah, did not have any rain for five years, between 1987–92 (fieldwork notes).

[7]   For a political profile of shaykh Abdullah see Koszinowski, T. (1993) Abdallah Ibn Hussain al-Ahmar. *Orient*, 34, 335–41.

[8]   For another example of royal lands aquired by a powerful shaykh see Dresch (1993:380).

[9]   *iktisab* (Arabic): acquisition of property or rights by uninterrupted possession of them for a certain period (Wehr 1976).

[10]   See a recent article on Yemen in the *National Geographic* magazine (April 2000) where Uthman is introduced as 'elected shaikh by the tribal elders only a year before, the handsome,

28–year-old has gained high repute for the fairness and rapidity of his judgements, so that among the throng are members of other tribes, drawn by his burgeoning reputation.' (page 36)

[11] According to one local source Awjari receives SR 50–75,000 every three months (US$ 5000–6000 per month). The other leading shaykhs in the area are said to get SR 18,000 every three months (US$ 1,600 per month). Exchange rate: Saudi Riyal 3.75 = 1 Dollar. Earlier, 'In 1983 certain shaykhs in the north-east were getting YR 15,000 or YR 20,000 [with a value of US$ 3,225–5,479 at the time] a month from Sana'a and as much from Saudi Arabia.' (Dresch 1993:19).

[12] Gingrich (1993:263) says that 'Wa'ilah have an old and famous reputation for never having accepted very much state influence, and having paid no taxes.'

[13] One of the sons of Qa'id Shawit's brother, a military officer, was killed for blood revenge during the 10–year-long Bani Uwayr – Sufyan conflict while returning one evening from a military camp. Qa'id Shawit is also one of the three men implicated in the assassination of prince Abdullah in 1969. The other two in the trio are thought to be Salih Hindi of Al Ammar and Ibn Haydar of Al Wasit (Harf Sufyan).The prince was on his way to the Friday prayers at the al-Hadi mosque in Sa'dah when he was shot near the Sa'dah home of Qa'id Shawit.

[14] The killing of women in tribal revenge is perceived as dishonouring in tribal society.

[15] see also Yemen Times 35 (30 August 199) where Suq al-Talh is described as the 'most famous market' for weapons 'at which all kinds of arms are available from pistols to heavy guns, rocket launchers, spare parts of tanks and wireless communications.'

[16] The *mafraj* of one trader had the capacity to seat 140 people.

4 Factors and Perceptions Influencing Expansion of Irrigated Agriculture

ardak – 'ardak
your land is your honour (Arab proverb)

This chapter examines the socio-economic and political forces that acted as a catalyst to groundwater exploitation by modern tube-well technology. A central part in this discussion will be the extent to which tribal perceptions vis-à-vis government politics triggered the acceleration of groundwater-irrigated agriculture. Hydrogeological data and water assessment studies of the region will be analysed and explained in the context of socio-economic and socio-political developments and changes that affected the area between the mid-1970s till the 1990s. It is argued that environmental degradation, and especially the unsustainable mining of the Sa'dah basin's groundwater resources from the mid-1970s until the mid-1980s, can be explained as the outcome of unequal power relations, political interests, and the changing ability of actors to control or resist other actors. It will be shown that within Sa'dah's politicised environment notions of political autonomy and resistance to state control have remained core values and objectives of many actors and tribal communities. Attempts by the state to establish control over the Sa'dah region have had direct repercussions on groundwater development.

A basic framework for the analysis of this chapter is provided in **Figure 4.1**. It presents a schematic overview of the factors that influenced groundwater abstraction and expansion of irrigated agriculture.

Groundwater and Conflict Resolution

In the context of the Sa'dah basin groundwater development has, paradoxically perhaps, been both the cause and the result of conflict. **Figure 4.2** seeks to capture this process schematically. Issues of conflict and conflict resolution have moved from water to land disputes. Until the mid-1970s water scarcity was at the heart of most tribal conflict. Resulting feuds were often drawn-out, and their resolution was being difficult and costly. Once the issue over runoff rights had been settled however, groundwater exploitation offered a way to avoid potential conflict. Moreover, groundwater appeared to provide a new measure of freedom, independence and certainty beyond the reach of customary and Islamic water rights (Beck 1990:29).

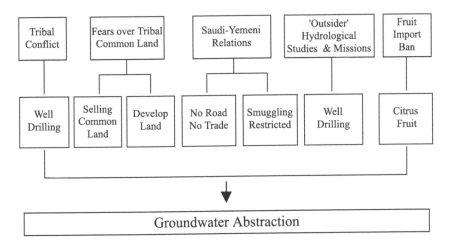

Figure 4.1   Socio-political and socio-economic factors influencing expansion of groundwater irrigated agriculture in the Sa'dah basin of Yemen

During the 1970s, Yemeni migrant workers in Saudi Arabia were exposed to technological possibilities in respect to groundwater availability and utilisation. In addition, those staying at home also witnessed with great surprise the new and unexpected groundwater 'miracle' when an the Italian company Furlani, building a new road through the Sa'dah basin in the late 1970s, drilled a deep-well to supply its own water needs.

Remittances provided the cash for drilling and equipment was brought in cheaply and tax-free from across the border. Privately owned wells promised secure water supplies, a greater amount of autonomy and a permanent resolution to potential conflict over the resource. At the time, people were certainly not aware of the long-term impact – the perception simply was that God had mercifully rewarded them with the 'gift' of water as he had blessed the Saudis with the 'gift' of oil. However, what was perceived as a recipe for conflict resolution over water now turned into potential conflict over land rights. A number of socio-political factors and perceptions were responsible for this development. As land values soared during the late 1970s, and especially as a result of the 1984 fruit import ban, claims over land have led to renewed feuds, disputes and drawn-out conflicts. Whereas the tribal shaykh's ability to mediate and solve conflicts used to rest primarily on his social standing and on his knowledge of customary law, his political influence and power at the national level now increasingly determine the outcome of such conflicts.

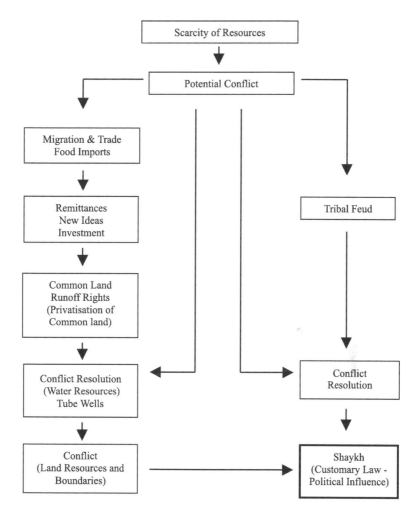

Figure 4.2   Conflict resolution: from water to land disputes

Fears over Loss of Land

San'a – Sa'dah Road

The road from San'a to Sa'dah, passing through the central highland terrain of *bilad hashid wa bakil*, has been travelled by pilgrims, traders, local tribes and others since time immemorial (al-Thenayian 1997:244). Before the completion of the new road in 1979 (Meyer 1986:265) the 245 km journey from the capital to Sa'dah took about eight to ten hours and was possible only by four wheel-drive vehicle or truck. The improved road cut down travelling time to four hours and opened-up all kinds of new opportunities for business and trade. It also shifted control and power between tribal actors as will become apparent shortly.

After passing through the arid region of al-Amashiyyah that appears to be devoid of human settlements, the road finally climbs up and reaches Suq Al Ammar, the tribal market of the Al Ammar tribe. Villages and small hamlets come into view, an indication that Sa'dah is only another 35 km away. Moments later one passes the large home of shaykh Hindi of Al Ammar, then the new road curves and enters al-Mahadhir (Sahar) with is prosperous market village of al-Qabil before descending into the Sa'dah basin from the west. A look at a 1974 map[1] reveals that this last leg of the journey diverges from the course of the road travelled on over the past millennium. The old road takes an easterly direction at the point where the tribal boundaries of Al Ammar (Hamdan al-Sham) and al-Mahadhir (Sahar) meet. From here the old road to Sa'dah used to pass through the tribal territory of the Wada'ah, who are part of Hamdan al-Sham.

In the mid-1970s, when the rough track was to be improved, the Wada'ah were worried about the consequences this might have for them. Not only were there fears that some of their land would be appropriated in the name of 'national development', but they were also concerned about tribal liabilities and safety aspects. A tribe is responsible for violations that occur on its territory (Dresch 1993:334). Even today the San'a – Sa'dah road is frequently 'cut' (*qata al-tariq*) by tribal communities as a result of violations perpetrated by other tribes on the territory of the former. During field work for this book, for example, in January 1999, the road was blocked by Hashid tribes as a result of the kidnapping of the author's friends . A Bakil tribe had abducted the six foreigners from Hashid territory and therefore violated their territorial peace. Consequently, Hashid blocked the main road to control the movements of the Bakil tribes in question. The above story is intended only to help us appreciate Wada'ah's concerns at the time. To them tribal liabilities outweighed potential benefits. Consequently, the approach of the new road to Sa'dah is through al-Mahadhir (Sahar) and its market town al-Qabil which has boomed as a result of this development.

Al-Mahadhir represents a separate small plain collecting runoff from a large area, including the mountain plateau of Bani Uwayr. The road's new approach, built in the mid-1970s, cut across a number of runoff zones and altered the flow of rainwater. Moreover, the value of land along the new road increased dramatically and created incentives to develop the runoff areas. According to a number of well-connected informants, these developments, besides those already mentioned in the previous chapter, acted as a catalyst for the 1972 Ijri arbitration.

Saudi Road

Between 1978 and 1981 another road was built further north that had a considerable impact on groundwater development. Financed by Saudi Arabia and constructed by an Italian company the road was planned to connect the Sa'dah basin with the Saudi Arabian city of Dhahran.

Prospective benefits from trade and services-related activities with Saudi Arabia provided strong incentives for many from the land and water scarce highland plateau regions to migrate into the Sa'dah basin. They bought land along the asphalt road running north through the basin as far as the small administrative town of Baqim near the Yemeni-Saudi border area. Expectations were that once the link through the mountains with the Saudi town of Dharan al-Janoub was completed, landowners along the road would engage in booming trade and service opportunities sparked off by migration and remittances. However, only in 2001 was the road link finally completed. Large road signs just north of Sa'dah saying 'Dharan al-Janoub 115 km' (**Figure 4.3**) only serve as a reminder of what could have been much earlier. The hopes of many who had planned to invest in small economic ventures were dashed, and those who had bought plots along the road were left with the only alternative – irrigated agriculture and the production of cash crops, especially fruit.

It is presumptuous to claim understanding about the political dynamics that determined the 'end' of the road at the time. The views and perceptions of informants, which include tribal-political personalities, suggest that competing interests and tribal rivalries were at play. The bulk of cross-border trade, whether labelled official or smuggled, was controlled by the Hamdan tribes of Wa'ilah on the Yemeni side and their allies across in Najran. Inevitably, the new road link would have shifted some of the control as well as some of the economic benefits to tribal populations further west – from Wa'ilah to Khawlan b. Amir. While these reasons might have affected the completion of the road it is also likely that the 60–year-old Saudi-Yemeni border dispute, now settled, and old Yemeni claims to Asir, Jizan and Najran provide plausible answers to this question.

What can be said, however, is that the Italian company appeared to have packed-up and left in a hurry. They were waiting for the last payments from the government in order to hand over the road camps. Their 'southern' camp was located on former *waqf* land. According to some, Faysal Surabi, the uncle of the present shaykh of the Bani Mu'adh came with 200 men of his tribe and demanded the hand-over of the camp to them. The company had no option but to comply, which indicates where the real power was centred as late as the early 1980s. In response, the San'a government

Figure 4.3   Road sign near Suq al-Talh
Dharan al Janoub, 95 kms north of Sa'dah is a town within Saudi
territory. Until 2001 the road, built in the late 1970s, abruptly stopped
just beyond the Yemeni town of Baqim, a short distance from the
present border. For this reason, perhaps, someone has tried to cross ou
the name. Hopes were for the crossing to be officially opened in 2002.

refused to release outstanding payments arguing that the company had failed to hand
over the camp to the government.

Present claims of ownership as well as land use of the former camp area are
informative and relevant to our understanding of Sa'dah's politicised environment.
After first appropriating the land, the shaykh's nephew and successor apparently
now leases the land back to be used by the San'a government. The government's
guesthouse *(dar al-diyafah)* has been built on the site and is used to host 'important'
government officials or guests of the government such as ambassadors. The
president stays there when he visits Sa'dah, which is not very often.

The building of the road, however, had other interesting side effects. It created
awareness of and shaped perceptions vis-à-vis groundwater availability. At the start
of operations the Italian road company drilled a deep tube well at the village of al-
Darb, a short distance north of Sa'dah along the proposed road, where the foreign
experts *(khubara)* had a camp. This had a tremendous effect on the people of the
immediate area. The impact this event had on local people from various places and
communities was repeatedly confirmed to the author – 'before this event many were

unaware of the water resources resting undisturbed in the ground. This gave people the idea to drill for water and to develop the land.' However, as for the road and the anticipated development of service opportunities, many who had hoped to diversify their livelihoods away from agriculture remained stuck with it.

Subsequent road developments through the tribal areas of 'Upper Yemen' have often developed into similar and highly contentious issues over control and access. The construction, in the 1990s, of a road linking the town of Huth, 120 km south of Sa'dah, with the lowlands of the Red Sea resulted in the kidnapping of foreigners by Bakil tribes. They felt that Hashid, with the help of their politically well connected shaykh, Abdullah b. Husayn al-Ahmar, had manipulated the final direction of the road so as to circumvent and avoid Bakil territory. The result would have been freedom of movement for Hashid while giving them control over the movements of Bakil tribes, a fact Bakil sought to correct by putting pressure on the government through the kidnapping of the foreign tourists.

Resource Capture: al-Maqash – Al Khamis

The village of al-Maqash has already been mentioned in the context of surface water runoff rights. It will be remembered that they are a Hamdan tribe from the east who, for reasons related to blood feud, settled among the Sahar tribes. The village community itself has, over the past decades, been plagued by numerous rivalries resulting in serious shoot-outs, evidence of which can be seen from the main road. Bullet marks are left in the mud-structure of their tower-shaped homes, especially around the upper windows. The exact details of these conflicts are not always clear but one thing is certain, at least to the tribal people around them, – their fighting is over land, and by implication, water resources.

A look a **Figure 4.4** reveals that their geographical location has political as well as strategic significance. The area of al-Maqash is at the southern edge of Sa'dah airport. To the north of al-Maqash lies the village of Al Khamis, which is part of Bani Mu'adh (Sahar). In 1980 the airport was only a small landing strip without a fence or a wall around it. However, at the time, the presence of the government and the military became increasingly felt. Rumours spread fears that uncultivated common land was at stake. In response communities sought to secure their land resources by privatising these areas.

In the course of these developments a quarrel ensued over the runoff rights (*sabb wa salab*) of a sizeable grazing area (*mahjar*) between the village of al-Maqash and their up-stream neighbours. Al-Maqash claimed both the right to the land and the runoff (*sabb wa salab*) while Al Khamis maintained that the land of the *mahjar* (*salab*) belonged to them. As the dispute remained unresolved government forces seized the opportunity to appropriate the grazing ground (about 60 hectares). The army moved in with weapons and sealed off the area with a large mud wall. Surprisingly, al-Maqash, who had claimed the lion's share, stayed quiet and did not confront the intruders. This was explained by rumours of a substantial transfer of

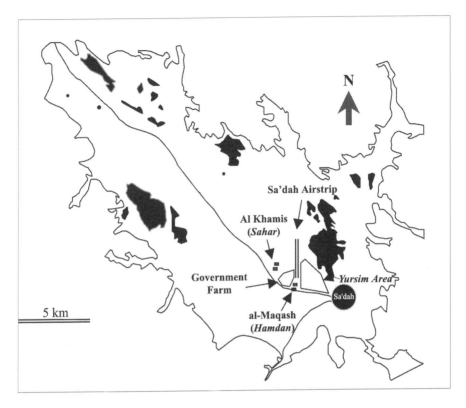

Figure 4.4   Resource capture

money and power to al-Maqash's leading political actor, shaykh, mediator, business man, and farmer.

The strategic and political benefits of appropriating land at the periphery of an airstrip seem obvious. But there might have been other considerations at play. The man in question is the son of a Wa'ilah shaykh[2] and an MP (House of Representatives) for the Kitaf (Wa'ilah) constituency. It serves us well to be reminded that the Wa'ilah, although inhabiting an environment which is characterised by its lack of natural resources, control the main traffic of cross-border trade. All other tribes doing 'business' across the border depend on them for safe passage. The shaykh is a 'big trader' (*tajir kabir*) and has benefited greatly from his own tribal connections and from the quest for power and control by other actors. During the early 1980s, when the San'a government sought to control cross-border trade, it only seemed politically prudent to co-opt the services of such an influential personality, be it at a high price.

Al-Maqash's main actor has done well. He owns a large house and other properties near al-Maqash and just along the main road through the basin. From there his farm stretches to the east and north. His trading activities involve – *Allahu a'lam* (God knows best) but certainly include the selling of irrigation equipment. Reports that he had recently installed drip irrigation in his large citrus orchards were found to be untrue. Politically, he is a representative for the district of Kitaf, in the heart of Wa'ilah country. A few years ago, when someone from Kitaf had his car stolen in the Sa'dah area, his services were enlisted. The car was soon retrieved from al-Buq, the Yemeni-Saudi border crossing point.

The land seized by the government has, for the time being, become the 'model' farm of the Ministry of Agriculture (**Figure 4.5**). Fruit (mainly oranges and apples) are produced on half of the land. For the purpose of our discussion here a few observations about land use of the remaining half of the government's farm should be included. During 1998 a dairy project was established in the corner closest to the airport. In a modern barn 40 Dutch cows are kept to produce fresh milk for the emerging supermarkets in Sa'dah town. If demand can be encouraged the dairy operation can be extended up to a 140 cows. The pasteurisation process is carried out at an adjacent building and with high-tech equipment that is state-of-the art technology by Yemeni standards. Some of the fodder is trucked all the way from the Tihamah. Significantly, the majority of the alfalfa needed is produced on the government's model farm, designed to exemplify water use efficiency. In this case, by contrast, the government's tube wells ensure sufficient water supply for irrigating alfalfa, which has the highest water requirement of all the crops grown in the area. What is revealing in the context of power and control over resources, the project described above is said to be owned by the deputy military commander in charge of the Sa'dah province. He is a half brother of a leading Talh shaykh and MP, who is frequently seen in company of the president.

At present, there are no regular flights from Sa'dah's airport. Depending on future political developments and, especially, on future relations with Saudi Arabia, the political and economic value of the land around the airstrip could increase enormously. In that case the future of the dairy operation appears uncertain. The dairy establishment, though, could provide a convenient argument for land rights established through prior use, especially, that is, for those with the right connections. Reports of conflicts resulting from resource capture make frequent headlines in many Yemeni newspapers and the notion *al-ard - 'ard* – 'land is honour' determines tribal perceptions vis-à-vis the government.

Yursim Area

In contrast to the fate of the *mahjar* just described the outcome was quite different for the area of Yursim, to the south east of the Sa'dah airport.[3] Lessons had been learned

Figure 4.5   **(A) The government's model farm at al-Maqash** (top photo)
Pipes for drip irrigation are in place but are used only occasionally.
(B) The village of al-Maqash in the background (bottom photo)

the hard way, by losing valuable land. Moreover, the respective parties felt humiliated since they were unable to defend their land and, by implication, their honour.

In 1982 the San'a government extended the small Sa'dah airport. Areas around the airstrip were sealed off and the landing strip was widened. In the immediate vicinity of the airport communities with common land resources were alarmed. The al-Maqash and Bani Mu'adh acted quickly and privatised the *mahjar* area known as Yursim. Parties with claims to rights (*sabb wa salab*) were aware that delays to privatisation, caused by endless disputes over the runoff land, might play in to the hand of skilful negotiators and/or third parties. Compromises were made and the land became private property in 1982, the same year that the airport was extended.

By the late 1980s all available land in Yursim had been developed into irrigated farms. A drought in the eastern Wa'ilah areas, that lasted from 1980 -1987, acted as a catalyst for many Hamdan people to invest in land and groundwater irrigation in the Yursim area. There were also other incentives and factors that acted as a catalyst for migrating into the Sa'dah basin, as will be shown in a subsequent section on migration. Asking about the present land value in the Yursim area the answer was- 'it is priceless, there is no more *ard bayda* (white land, i.e. uncultivated) for sale, nor will anyone here sell any'. This statement is easily confirmed by taking a view from the flat roof of the informant's house. In the distance the runway of Sa'dah's airport lies silently, symbolic perhaps of both the government's ability and limitation to control the outcome of events in Sa'dah's politicised environment. A map of agricultural land in the Sa'dah basin based on 1980 air-photographs shows no evidence of agricultural activity in the Yursim area (Danikh and Van der Gun 1985:Fig.11). However, a well survey conducted during 1982/83 indicates the presence of ten wells in the area, seven of which had been drilled in 1982. Only one well was from 1978 with no date of construction for the remaining two (Gamal et al., 1985: 65f). **Figure 4.6** shows the area in 1998, Yursim has become one of the most densely irrigated areas in the Sa'dah basin dotted with large-size farms that are owned by people from al-Ahnum (famous for its qat production) and lower Khawlan to tribal folk from the eastern Bakil tribes. It is also an area where groundwater levels declined by 50 to 55 metres in the period from 1983–1992, a drop of six metres per year (DHV 1992:44 and Fig.4.20).

Hamazat Area

Northeast of *Yursim* but on the other site of the rocky outcrop is an area called Hamazat. The people of Hamazat are exclusively *sayyids* (Heiss 1987:63). Placed between the Sahar to the west and Hamdan to the east Hamazat is *hijrah* (protected enclave) to both and under the protection of both. However, they feel closer to Hamdan and their shaykh, Abdullah al-Awjari, who it will be recalled, has his large farm immediately south of Hamazat. Like the shaykh's father, the Hamazat *sayyids* fought for the Imam. As a result some spent many years in Najran, Saudi Arabia.

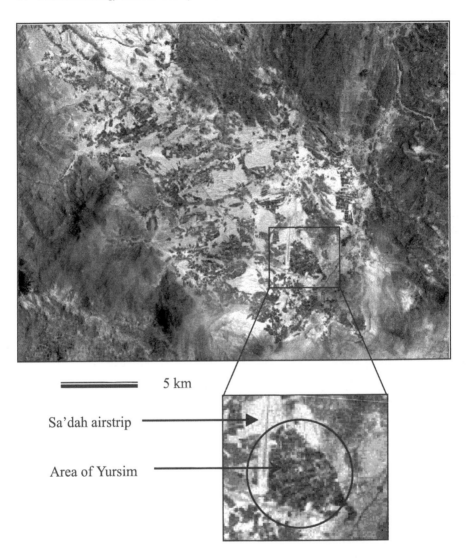

5 km

Sa'dah airstrip

Area of Yursim

Figure 4.6 The area of Yursim
was privatised and subsequently developed after 1982 in response to fears that communal lands left undeveloped might be lost to government plans to expand the Sa'dah airstrip. Groundwater levels in Yursim dropped by over 40 metres between 1982 and 1992 (DHV 1992:45). Date of satellite image: August 1998.

For political, social and religious reasons the Hamazat *sayyid*s felt vulnerable to having their land taken from them. First, they had supported the Imam. Secondly, as a small group that had become marginalised after the revolution they lacked the muscle to defend themselves. Thirdly, the land they farm is considered *waqf* (religious endowment) and for reasons discussed in Chapter 3 much of the Sa'dah's *waqf* was contested by emerging actors (Gingrich and Heiss 1986:21).

Most importantly, as the president announced 'The year of Agriculture' (*am al-zira'ah*) in 1984 promoting food self-sufficiency, these men of religion could foresee the consequences this might have for their uncultivated common land. Long before the idea of homesteading lured settlers to tame the wild west of America, the *shari'ah* provided for barren land (*ard mawat* or *ard bayda*) to be appropriated by the Muslim ruler. Alternatively anyone from the Muslim community bringing *ard mawat* into production for a period of three years could claim ownership of it (Maktari 1971:1ff). Whereas in the *shari'ah ard bayda* is owned by the government in customary law it belongs to the tribe. Perceptions of *shari'ah* and customary tribal law seem to be at odds with one another. The Sudan provides just one vivid example where, under the same premise, the state has appropriated large tribal territories. Land has been handed out as gifts and rewards to military officials and those with the capital to bring it to 'life' (*ihya al-ard*) (Suliman 1993:59–62).

In the Sa'dah basin, as most other places, it is the perceptions of rather than the fundamentals of Islamic jurisprudence that determine people's actions. Perceptions were not only shaped by realities and events in the Sa'dah area but also by the stories migrants brought back from Saudi Arabia. For many who had spent long periods working across the border, fears that a strong San'a government could eventually take tribal lands were based on the belief that the Saudi government had done just that to their tribal communities.

During the course of 1984, the year of Agriculture (*am al-zira'ah*), the Hamazat privatised their *mahjar* (runoff area). A majority owned only small plots while others could claim no private land (*mal*) but owned livestock and depended on the runoff for grazing. About 50,000 *hablah* (125 hectares) of runoff (*mahjar*) were to be divided according to the number of males, including the male children. In the case of Hamazat the 1972 Ijri solution, described in Chapter 3, presented a slight problem. Some households had plots irrigated by runoff. They could claim vast runoff areas as their right. Others irrigated their small gardens from shallow hand-dug wells in the village and could claim only small shares of the common land. Those primarily occupied with religion as well as those herding livestock were also at a disadvantage. Moreover, the situation was made worse by the fact that those households with the smallest entitlements had the largest number of children. When this inequitable outcome resulted in jealousy and even armed conflicts, the community decided to review the situation. It was eventually agreed that their large runoff area be divided-up equally between the total number of males, disregarding the customary rights of *sabb wa salab*. According to the revised formula 'every male, including Ali, who was born that same day, received about 100 *hablah* (0.25 hectare) land'.

The Hamazat case illustrates the adaptive capacity of this particular community. Religious and customary entitlements were adjusted to serve the common good and to ensure a more equitable outcome for all. Moreover, the *sayyid* community quickly solved their internal conflict in an attempt to prevent resource capture by outsiders. Unlike many other communities, who sold parts of their land to people from outside their immediate tribal communities, the Hamazat have prevented 'outsiders', in particular non-*sayyid*s, from settling among them.

Hydrological Missions

In 1982 the Ministry of Oil and Mineral Resources in association with Dutch hydrologists identified activities for training Yemeni hydrologists. A number of areas, representing the country's different climatic and hydrological regimes, were selected to carry out the various water assessment studies. The Sa'dah basin was chosen because of the lack of hydrological data on the area. The exercise was meant for capacity building and to train Yemeni hydrologists. Thus the field studies were seen as an excellent opportunity to provide on-the-job training for promising candidates. Activities in the Sa'dah area included an assessment of groundwater availability and a well inventory. Fieldwork lasted from November 1982 until August 1983 and involved a considerate number of Yemeni and foreign personnel, about 40 and 15 respectably (Van der Gun 1985:1;51f).

Apart from a number of the area's leading shaykhs, who had been informed about the hydrological teams, the majority of the farmers claim that they had no idea about the mission's objectives. As a consequence, outcomes were quite different from intentions. Perceptions varied according to locations. Nearer to Sa'dah town, where government presence was increasingly being felt, many farmers perceived these hydrological teams as extended arms of the national government.[4] They feared that new legislation to control groundwater abstraction rates and irrigated agriculture was imminent and could prevent them from developing their own fallow land.

Following the euphoric news in 1984 that 'large' quantities of oil had been discovered in the Ma'rib area other farmers thought that the purpose of the well count in the Sa'dah basin was to determine the allocation of diesel fuel to operate the pumps. The belief was that those who had a well or could show that a well was being drilled could get a permission to buy diesel fuel at special subsidised rates.

One farmer recalled that even after a number of repeated visits by these foreign and local experts (*khubara*) he still had no idea about the purpose and result of their investigations. '*Your water is not suitable for drinking – al-ma haqqak ghayr salihah li l-shurb*' is all he was told.

Suspicion nourished by a lack of information and hopes for cheap subsidies triggered similar responses – more wells were dug. It was observed by members of the hydrological team that soon after surveying particular areas farmers would start to take 'precautionary' measures. Earthen walls were bulldozed around landholdings of previously fallow land, ground would be levelled and prepared, and lots of new wells were drilled. In many cases it was believed that land not agriculturally used could otherwise be lost in the contest for control over the natural resources of the area.

Saudi-Yemeni Relations

The clans prospered most in time of political instability,
when the crown was least able to exert control.
(Source: Clan feuds in a land apart, TIC, Fort William)

Smuggling

External remittances earned by Yemeni migrants in the late 1970s led to unprecedented changes affecting agricultural production, spending patterns, and social status perception (Weir 1985). Much of the newly earned riches were spent on vehicles and consumer items which were smuggled across the border into the Sa'dah area where the goods were sold at large tribal markets outside government control (Burrowes 1995:337). With an estimated value of over one billion US dollars, smuggled goods equalled official imports in the 1970s (Quist 1990) allowing Burrowes (1995:337–338) to conclude that 'smuggling is a major feature of Yemeni economic life'. His statement is certainly correct with regard to the Sa'dah region. The subject of cross-border trade is nothing to be ashamed of and will naturally come up in almost every discussion with individuals and groups from the Sa'dah basin. From the mid-1970s until the early 1980s at least one third and possibly one half of Sa'dah's adult male population were involved (many still are) in the lucrative cross-border trade. Almost anything could be smuggled, from cars and cranes to cornflakes. Mountainous terrain presented the only limitations. For example, **Figure 4.7**, a photo taken by the author in 1986, shows the result of an attempt to navigate a drilling rig across a mountain range near the Saudi-Yemeni frontier.

Local people are also quite adamant that the basin's powerful and influential traders of today, almost without exception, made their fortunes from smuggling. Profits lured many shaykhs and others from outside the immediate area into the basin. Even for all those who only carved out a small slice of the pie, incomes from smuggling, and not agriculture, provided the main part of their livelihood. This conclusion, drawn from countless interviews and discussions, is also confirmed by a 1993 survey which mentions that border trade was the main occupation of many people in the area (DHV*a* 1993:15).

The volume and nature of cross-border trade is symbolic of Sa'dah's politicised environment. As Borrowes (1995:337) perceptively remarks, 'smuggling has been indicative of the extent to which the state has been unable to control the periphery

Figure 4.7   Drilling rig
An attempt to smuggle a drilling rig from Saudi Arabia to the Jauf in 1986 came to a temporary halt when the truck turned on its side on a narrow mountain pass leading up to Barat (Bakil).

and its borders.' In the early 1980s the government of the then Arab Republic of Yemen mounted several attempts to control activities along the border (Burrowes 1995:337). However, the impact of these measurers was often short-lived. Powerful personalities in Saudi Arabia and tribal leaders in Yemen derived huge financial and political benefits from smuggling. Their social and political links and networks made it difficult to control cross-border activities (Burrowes 1995:337). Moreover, other tribal factors would appear to make effective and permanent control highly unlikely. Firstly, the deputy commander in charge of Sa'dah's armed forces is from the area. Secondly, Sa'dah's tribal leaders have been successful in positioning many of their tribesmen into various army units. Bani Uwayr (Sahar) and Barat (Bakil) are only two examples.

Attempts in the early 1980s to control the unofficial traffic of goods between the two countries did not affect some of the main actors in this domain for another reason. By that time many 'big' traders were no longer dependent on smuggling. Instead, contacts and benefactors in 'higher' places helped them to import officially with similar financial results. Many of the smaller local traders that were forced out of smuggling during the time maintained that measures taken by the army only shifted further the loci of power to those with political links and established networks.

As a consequence, smuggling activities became too risky for many of the average traders. When the government's ban on the import of fruit and vegetables in 1984 provided new economic incentives and raised hopes for a livelihood in agriculture many that had worked in smuggling shifted to a future in agriculture instead.

Changes in Saudi Arabia

From the mid-1970s and up to the early 1980s between one third and up to one half of Sa'dah's adult male population worked in Saudi Arabia. Journeys back home to Sa'dah were made once or twice a year, but especially for Ramadhan. The Muslim obligation of pilgrimage to Mecca (*hajj*) 40 days after the end of Ramadhan provided a good reason to return. Birth figures in hospitals support these patterns of migration. The number of deliveries nine month after Ramadhan rose dramatically.

The season for the grape and sorghum harvests in late summer and autumn were also favoured occasions to visit the basin. These harvest times also helped to strengthen and renew cultural identities.

In the late 1970s, and especially as the Islamic month of Ramadhan approached, the volume of cars pouring in to the Sa'dah basin was striking. Most were overloaded with consumer goods and foodstuffs from Saudi Arabia. Striking, too, was the increase in fruit available in tribal and town markets and along many of the country's highways. Large quantities of oranges and apples of US and Lebanese origin were smuggled from Saudi Arabia, especially for the festive seasons of Ramadhan.

As evident in the precious chapter, Sa'dah has always been famed for its grapes. As consumer patterns changed and fruit, especially oranges and apples, became a lucrative cash crop, farmers in the Sa'dah basin were eager to experiment with fruit production.

The majority of Yemeni migrants in Saudi Arabia in the 1970s, over 50 percent according to Meyer (1986:46), were employed in the building industry. But in discussions with individuals and groups from the Sa'dah basin it became evident that many others worked on agricultural farms in the Saudi Kingdom. Figures indicate a higher proportion of Yemeni migrants engaged in agricultural labour for the three former Yemeni provinces of Jizan, Asir and Najran than for the rest of the Saudi Kingdom (Meyer 1986:47). The pioneering work of Abdulfatah (1981), who describes the agricultural changes in Asir, mentions the investments in agriculture made by the traders of that region during the late 1970s.

Most Yemenis identify deeply with the cultivation of the soil (Dresch 1993:307; Caton 1990:32f; Swanson 1979:38). Exposure, whether through working direct on Saudi farms or by observing state-financed and state-subsidised agricultural development in the Kingdom, had a profound impact on Yemeni migrants and changed perceptions about the values of land and water, especially the use of groundwater.

As long as conditions for work across the border were favourable and as long as remittances remained high there were no immediate economic incentives to invest in agricultural production at home however.

The preferential conditions enjoyed by Yemeni migrant workers in the Saudi Kingdom started to change in the early 1980s. As we have discussed earlier, the Sa'dah region's communities share cultural and social characteristics with their immediate neighbours across the present political boundary. These factors have given Sa'dah's migrants a considerable advantage over migrants from other regions in Yemen with regard to doing 'business' in Saudi Arabia.

Up to the early 1980s Sa'dah migrants, as indeed Yemenis in general, had been allowed to establish, own and operate their own businesses. In fact, Meyer (1986:46) reports that as early as 1974 one out of eight Yemenis was either self-employed or had established his own building company employing primarily fellow Yemenis. By the early 1980s new regulations began to constrain the entrepreneurial spirit of Sa'dah's migrant population (Findlay 1994:114f; Morris 1986:302).[5] First, Yemenis no longer were permitted to run their own enterprises inside the Kingdom. Secondly, Yemenis were no longer able to drive their own, commercially registered, trucks (*mana'u alayhim al-naql*). This, in particular affected many who were in the business of transport and hauling. Thirdly, by the early 1980s the building boom in Saudi Arabia had peaked and as a result there were fewer opportunities for work. And with the increasing fluctuations and subsequent fall of oil prices the Saudi government preferred to bring in cheap labour from the countries of South Asia and the Far East.

Moreover, during the early 1980s a renewed assertiveness of Islamic belief and practice, sparked perhaps by events in Iran, Egypt and Sudan, was making itself felt. Muslims from Sa'dah follow the Zaydi school of Islamic interpretation (a Shia off-shot). Discriminating and derogative remarks, such as *zuyud* (Zaydi) and *makhnuth* (homosexual) were reasons enough for some to consider a permanent return to their home communities in the Sa'dah basin.

Economic conditions started to worsen at home too. The value of the Yemeni Riyal fell steadily after 1984. A number of leading traders and shaykhs had made fortunes during the peak years of cross-border trade, between 1975–1980. With increasing economic and political uncertainty they now felt that it was 'better to have land than to have money' a remark made by one leading tribal actor who saw the value of his land in the capital sky rocket, from YR 20,000 to YR 90 million ($US 720,000) between 1986 and 1996. In the Sa'dah basin too, for reasons of social prestige and political image acquisition of land became important as many sought to reassert Yemeni cultural values from the mid-1980s onward.

The production of citrus and apples, especially, was actively promoted in the Sa'dah basin in the wake of the government's ban on the import of fruit and vegetables in 1984. Consequently, well drilling increased dramatically during the following three years as the entrepreneurial tribal population responded to economic signals and potential market opportunities. Traders and shaykhs were also the first to benefit from government subsidies and gifts. They set trends and their communities soon followed in the production of potentially lucrative 'new' crops, mainly citrus and apples.

Fruit Import Ban

the traditional relationship between the land and the people was broken with many clan chiefs choosing the role of landlord before leader.
(TIC, Fort William)

Figure 4.8 indicates the tremendous increase of tube wells following the government import ban in 1984 on fruit and vegetables. While the number of new tube wells drilled in the Sa'dah basin had fallen from 190 in 1982 to 84 in 1983 figures climbed to 274 in 1984 reaching an all-time high in 1986 when 287 new wells were drilled into the Sa'dah aquifer. The fruit import ban provided economic incentives to invest in agriculture. Trade, a main source of income for many – at least every other household had a member involved in cross-border trade – had become a risky business and proved less profitable. Those who had made money from migrant labour and/or trade earlier sought to invest their capital. The mid-1980s marked a new phase in groundwater irrigated farming as many new farms, especially citrus and apple orchards, were established in the Sa'dah basin. By that time the area's main actors, shaykhs and traders, had well established links with central government. They benefited in numerous ways from the government's agricultural policy, they alone had the capital needed to establish large farms, exceeding 5 hectares. The first to respond were two leading shaykhs, one from Sahar (Khawlan b. Amir), the other from Wa'ilah (Hamdan).

Shaykhs and Traders

The ascent to power of Bani Mu'adh's tribal leader, as outlined earlier, came in the mid-1980s. His large 25 hectare citrus orchard, with about 10,000 trees, was established around 1986, the same year the San'a government acknowledged his sphere of control by drilling a number of wells for him.

Shaykh Abdullah of Wa'ilah too, 'holds the government in his hand (*al-hukumah fi yadhu*)'. This phrase, related with an attitude of pride by men of the shaykh's tribe has to be understood in the context of the control naturally bestowed upon his tribe by virtue of their geographical location along the Yemeni-Saudi boundary.

On 62.5 hectares of land, said to be partly *waqf*, which was contested by various tribal leaders, the shaykh now irrigates about 20,000 citrus trees. Local people recall, that in 1984, when the government started to promote the production of citrus, the shaykh just went to where the Ministry of Agriculture stored the young seedlings and took enough seedlings to establish 25 hectares of citrus orchard. In

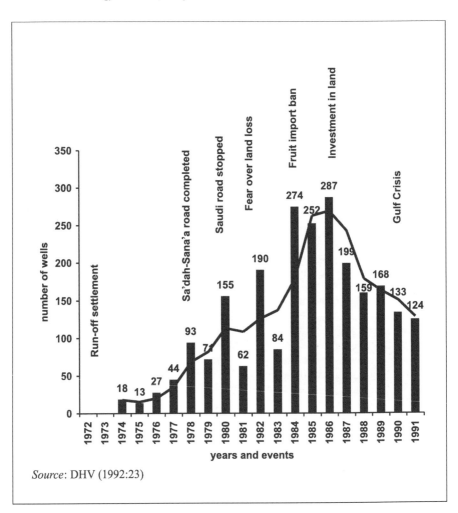

Source: DHV (1992:23)

Figure 4.8   Well development in the Sa'dah basin and political and economic events

The well numbers are extrapolated from the detailed well age data of 556 wells in 1992 among a total of 2457 wells at that time. The trend line is significant as it indicates that well drilling peaked in the years following the fruit import ban.

1996 the shaykh doubled his citrus orchard from 10,000 – 20,000 *hablah* (from 25 - 50 hectares).

The large fruit orchards owned by many other influential personalities, including Sa'dah's trading clans were all established following the fruit import ban.

The development of large fruit orchards by shaykhs, traders and those with the capital means sent important signals to those with fewer means. In contrast, the average Sa'dah farmer lacked the means as well as the connections to extensify agricultural production. Besides, to the smaller subsistence farmers there were other considerations and risks and most farmers started small, experimenting with results and watching trends. Prior to 1984, the year pronounced 'Year for Agriculture' (*am al-zira'ah*) by President Ali Abdullah Salih, fruit smuggled across from Saudi Arabia was cheap and there was little incentive to invest heavily in the production of fruit, considering that returns could be expected only after three to five years after planting.

Moreover, unlike some of the shaykhs and traders, who received large quantities of seedlings as political gifts, the number of trees the average farmer could buy depended on his income and farm size. Husayn, a young progressive farmer put his name down for 100 trees but received only 50. Also, in contrast to the 'haves' who opted for citrus, the average farmer preferred trees with a shorter maturing period, such as apples and peaches. Grape production was not affected since the quality of Sa'dah's vines was deemed superior by far to foreign imports. Raisins also remained an important subsistence food in the area.

However, the government's renewed emphasis on food self-sufficiency had other repercussions that negatively effected Sa'dah's groundwater balance. As discussed earlier, the 1972 arbitration over runoff rights had paved the way to privatise tribal common lands. Only by 1986 could it be said that many of Sa'dah's tribal grazing areas were divided and allocated to those with claims. What accelerated this process? First, there was a general perception that unless Sa'dah's farmers strengthened agricultural production the government would take up the task to cultivate 'white areas' (*ard bayda*). As indicated earlier, the *shari'ah* provides some rationale for this move and promotes the 'bringing to life' of 'dead' land (*ihya al-mawat*).

Secondly, tribal shaykhs and influential traders were increasingly co-opted into central government politics. As these shaykhs had considerable power over allocation and management of their tribal common land their people were concerned about the fate of these areas.

Thirdly, the 1984 import ban created demand for land, which led to an enormous increase in the value for land. In areas close to Sa'dah town land prices had already seen an increase of ten fold between 1975 and 1983, to YR 11,000 per *hablah* (YR 440 per m[2]) due to the expansion of the city westwards. Whereas earlier, in 1975, shaykh Fa'id had unsuccessfully offered people a *hablah* for one Maria Theresa thaler coin (one Yemeni Riyal at the time)[6] he gained great fortunes during that period as he controlled large sections of the said area.

Now, over the next decade, between the early 1980s and the mid-1990s, the value for potential land for agriculture increased to unprecedented levels, even in areas further away from towns and market centres. The price for one *hablah* (25m²) of *ard bayda* (white, i.e. barren land) rose from YR 500 – 1,000 in 1982 to YR 2,000 - 6,000 by 1997. In the area of Yursim, discussed above, one *hablah*, were it to be available, would be no less than YR 7,000 now. Those who had bought land along the main road north through the basin saw values go up to YR 35,000 per *hablah*. The price for plots along the Sa'dah – San'a road, at least for the first 5 km outside of town are up to YR 100,000 at present.

Privatisation of tribal common land left Sa'dah farmers with more land than they could cultivate. To finance well drilling and pumping equipment landowners sold part of their land. As land values increased, incomes from the sale of extra land were considerable.

The buyers came not only from the Sa'dah basin. The majority of Yemeni migrant workers in Saudi Arabia passed through and stopped in the Sa'dah area on their journey to and fro. Many from the barren and treeless highland plateau areas, described by Kopp (1977:24) as a natural environment that appears most unsuitable for human settlement, aspired to purchase land in the Sa'dah basin. Individual land holdings in these rocky highland areas of Hashid often don't exceed 0.5 hectares. Hashid, especially, had been in the transport and distribution business of the goods smuggled across the border (Dresch 1993:308f; 381). As a result many of these tribal entrepreneurs had links and connections with those involved in cross-border trade from the Sa'dah basin. Moreover, some had gained considerable profits and were looking for land as a means of investment.

By the late 1980s the water stressed area of al-Dumayd, details of which will be found in Chapter 6, had seen a particular high influx of new land owners, often from outside the basin. These include people from the main tribal confederations Hashid and Bakil with their numerous subsections as well as from tribes of Khawlan b. Amir with their subsections Razih, Munabbih, al-Mahadhir, Khawlan and Juma'ah. Farmers there share no history of co-operation with their host communities. In fact, a primary reason for moving to the Sa'dah basin may have been to break free from the need to share and co-operate over the scarce and limited water and land resources in their highland home territory. As will be discussed in subsequent chapters, these factors complicate the issues of co-operation and control over water resources and also appear to inhibit the formation of local initiatives and user groups.

Figure 4.9 summarises some of the main other reasons and incentives, push and pull factors, for migrating into the Sa'dah basin. It will be noted that a prolonged period of drought in the eastern Wa'ilah territories during the early-1980s 'pushed' many of these semi-nomadic people to settle in the Sa'dah basin.

The fall in value of the Yemeni Riyal, that started in the mid-1980s and accelerated after the Gulf Crisis in 1990, had direct consequences on groundwater irrigated agriculture as it forced many traders out of business. 'Had it not been for the devaluation of the Riyal we would not have gone into farming' was an enlightening statement by a small trader who went into farming only after the Gulf

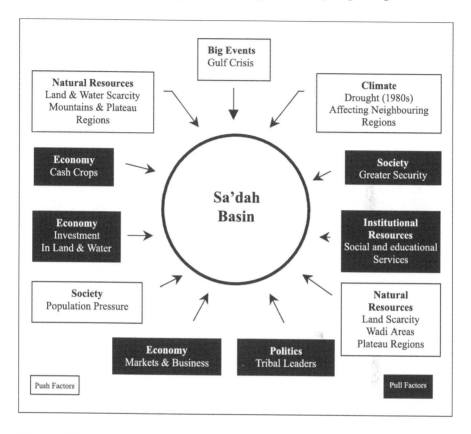

Figure 4.9   Migration into the Sa'dah basin: pull and push factors

Crisis. **Figure 4.10** shows the plunge of the Yemeni Riyal since the mid-1980s. Investments to drill wells and purchase pump and irrigation equipment became a way for many to 'save' their hard-earned savings from disappearing. At times the pace of inflation appeared alarming. One informant had 2.5 hectares of uncultivated land. In a desperate attempt to convert his savings into some form of value, he hurriedly decided to have a new well drilled for the future development of this land. By the time his well was completed the value of his remaining money had fallen to such level that he could no longer afford to purchase a pump.

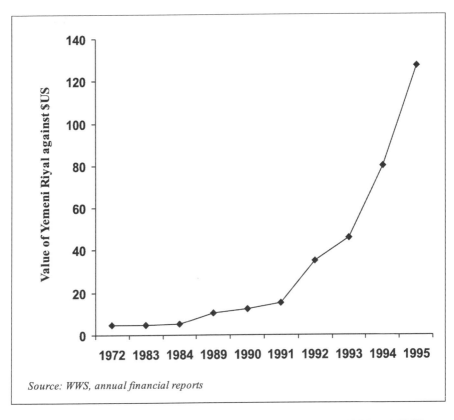

Figure 4.10 Devaluation of the Yemeni Riyal (YR) between 1972 and 1995

Water and Conflict

Although the volumes and goods of cross-border trade decreased from the mid-1980 onwards, San'a's import ban for fruit and vegetables in 1984 benefited Sa'dah's trading families as they represented the principal importers of pump and irrigation equipment. Moreover, Sa'dah's trading families owned the drilling rigs which had 'found their way across' from Saudi Arabia. With an estimated 45,000 tube wells in the country in 1994 (Yemen Times 22 May 1995; World Bank 1997:1; Kopp 2000:84) drilling was (and still is) a lucrative business. Drilling rigs owned by businesses from the Sa'dah basin drilled the length and breath of the country. Moreover, drilling provided jobs for people from the area, although these activities were, at times, highly political and involved risk to life and limb. The tragic accident of a friend vividly illustrates notions of resource capture, power and resistance.

In 1985, Muhammad was working on the drilling rig of his relative, one of Sa'dah's prominent traders, when he and the driver of the car were shot at. The driver died and Muhammad lost his arm. The San'a government had promised to drill a well for a community of the Hada tribe in Dhamar province. To no one's surprise, their shaykh determined the 'most suitable' location, which happened to be on land that he himself owned and where he had gardens. However, he died before the scheme was implemented and a dispute started between his two sons over who would become shaykh. Their rivalry split the tribe, one section following one son and the other his brother. Perhaps, in order to gain more support one son tried to have the well drilled on the land he inherited.

The group supporting his brother objected by arguing that the well was *maslahah ammah* not *maslahah khassah* (an Islamic principle denoting the common good over private benefits). Muhammad and his drilling team were caught in the middle of the conflict over location and therefore control of water resources as one of the shaykhly brothers sped to the provincial capital Dhamar in order to muster additional support from the government. The incident happened as Muhammad and his driver left the area soon after. Their vehicle was mistaken for the vehicle in which the one shaykh had just left for Dhamar.

They were shot at along the mountain tracks by the contesting section of the tribe. The driver died while Muhammad was seriously wounded. He was taken to hospital where he almost died had it not been for the fact that his employer and relative, one of Sa'dah's leading traders, chews qat (i.e. having close links) with the chief of the police force – the president's brother. Through his intervention Muhammad was flown out to Paris 'that very night' and later again to Germany for treatment.[7] What impact this tragic mistake had on the water supply for the disputed gardens of Hada's deceased shaykh is unclear. This tragic accident triggered another tribal conflict, now between the respective tribes from Hada (Dhamar province) and Sahar (Sa'dah province). It was finally settled only 3 years later, in 1988 when a group of Hada tribesmen travelled the 400 km north to the Sa'dah basin in order to slaughter beasts and pay compensation to the victim's family. By that time the Hada were involved in other disputes with tribes along the main road through *Bilad Hashid wa Bakil*. This meant that some had to make a long detour over territory considered safer, first down to the Read Sea cost, then north through the hot Tihamah until they reached the Saudi border, and finally eastwards, over the mountain terrain of Razih and Khawlan until the reached the Sa'dah basin. The celebration lasted several days and many shots were fired – this time in the air.

The above example shows the potential for tension and conflict that resulted from groundwater development. It also indicates ways in which different actors have benefited from these activities. Wells signalled important political gifts through which shaykhs were co-opted into power sharing and co-operation with the central government. Traders and business entrepreneurs benefited from working with both, shaykhs and national government.

Conclusions

Throughout this chapter it has become evident that perceptions of politics as well as notions of power and interests, control and resistance explain expansion and development of groundwater irrigated agriculture in Sa'dah's politicised environment.

In summary, **Figure 4.11** and **Table 4.1** capture chronologically the main developments and the socio-economic and socio-political forces and developments that facilitated the rapid and unsustainable expansion of irrigated agriculture in the Sa'dah basin.

In the early 1970s a number of issues led to the crucial arbitration over runoff rights, which subsequently facilitated the privatisation of many tribal grazing areas. Firstly, Yemeni migrants in Saudi Arabia were exposed to new ideas and possibilities in respect to groundwater availability and exploitation. Secondly, population increase at home had put pressure on the scarce surface water resources resulting in tribal conflicts over water resources. Groundwater development appeared to provide a way to avoid tribal conflict. At the same time it offered increasing measures of autonomy and freedom from the control of others.

From the late 1970s a number of socio-political factors and perceptions hastened the privatisation of tribal grazing lands. Firstly, there were fears that common land as well as uncultivated land might be claimed by the government seeking to gain political control over the area. As a result land was sold to neighbouring tribes who then migrated into the Sa'dah basin. Secondly, the development of a road to link the basin with Saudi Arabia triggered land sales.

Thirdly, hydrological missions initiated a variety of fears and responses. Fourthly, efforts to control smuggling pushed many into agriculture as their major source of livelihood.

Table 4.1   Sa'dah groundwater development: causes, consequences and constraints

Period	Forces	Constraints	Well Development
Until 1970			Hand-dug wells.
Early 1970s	Population pressure. Political fears. Remittances & technologies.	Communal land. Runoff rights.	Few deep wells. Diesel pumps on old wells.

1972	Arbitration over Runoff.	Labour shortage for agriculture.	Beginning of tube well development.
1978	San'a - Sa'dah road. Cash crops. Social change. Remittances. Relative low cost of tube wells.		Steady increase.
Early 1980s	Sa'dah - Saudi road. Changes in Saudi Arabia. Drought. Sa'dah water assessment studies. Fear over land loss. Land sales.	Tribal land disputes. Runoff rights.	Rapid increase.
1984	Fruit import ban. High returns (fruit). Returning migrants. Low cost for diesel fuel. Gov. campaign – food self-sufficiency.	Tribal land disputes. Runoff rights.	Dramatic increase.
1988	Investment in land. Devaluation of Yemeni Riyal. Falling oil price.	Tribal land disputes. Runoff rights.	Peak of groundwater development.
1990	Returning migrants. Inflation. Stop of trade and smuggle.	Economic cost. Awareness. Runoff rights preferred.	Tube development continuing but reduced level.
1994	Decrease of volume. Social status. Food self-sufficiency. Autonomy (war, crises). Livestock. Qat.	Economic cost. Capital. Low returns (fruit). Marketing.	Deepening of tube wells. Few new tube wells.

The import ban for fruit and vegetables in 1984 marks a further milestone for groundwater abstraction. External as well as internal forces provided additional incentives for the expansion of irrigated agriculture. Firstly, changes in the conditions for Yemeni migrant workers in the Saudi Kingdom persuaded many to

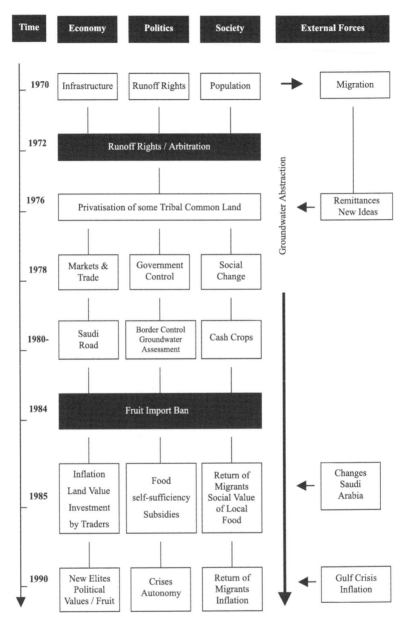

Figure 4.11   Sa'dah water development: causes and consequences

return home and invest their savings in the production of cash crops. Secondly, the devaluation of the Yemeni Riyal made cross-border trade more risky and less profitable. Consequently, many local traders and entrepreneurs started to invest their money in land. Lastly, the various crises during the 1990s (Gulf crisis, 1994 civil war, the Hunaysh dispute and tensions along the Saudi-Yemeni border) have all fostered a climate of uncertainty that reinforced tribal notions of autonomy and food self-sufficiency. In the next chapter, it will be demonstrated that current water use pattern reflect, to a large extent, Sa'dah's politicised environment.

Notes

[1]   The Yemen Arab Republic, 1:250,000, Produced for the YAR by the Director of Military Survey, Ministry of Defence, UK.

[2]   The man's father was shot dead only a few Ramadhans ago. Unarmed, he had left his house in the afternoon to buy meat from the near-by butcher shop on the main road when he was gunned down.

[3]   Yursam, according to al-Hamdani, describes one of the two groups that inhabited the town of Sa'dah at the time. The Yursam 'were a mixed group consisting of members of al-Kala' from the surroundings of Ibb, Hamdan and Khawlan (Heiss (1987:66).

[4]   The mission's principal hydrogeologist had the impression that 'in 1983 there was no government in Sa'dah. It is likely that the team [of hydrogeologists] were seen as the first advance of the government in to certain areas.'

[5]   The fieldwork of Morris (1986:302) in the early-1980s supports this: 'Longer-established migrants who have set up as building contractors, shopkeepers and restaurant owners are fearful now that a policy of Saudiisation is being extended to the service sector of the economy. Those Yemenis with neither Saudi nationality nor influential friends are being obliged to register in the name of a Saudi. ....As a result of these pressures, wages paid to labourers have fallen in real terms while those paid to skilled and semi-skilled construction workers have ceased to rise in the dramatic manner of the 1970s. Nimri labourers who use to earn eighty to a hundred Saudi Riyals a day were receiving only fifty or sixty by 1982.

[6]   These heavy silver coins are known as *Riyal fransi* or French Riyal. The were originally minted by the Hapsburgs and later by the Bank of England. In the nineteenth century these coins were used as currency throughout Arabia. Morris (1986:136) reports that in his study area in the western mountains *Riyal fransi* were continued to be exchanged in bridal payments until the 1970s. In 1997 a 'fransi' had a value of about YR 400. In the Sa'dah area these coins are still used today for bridal payments by certain social groups.

[7]   His upper arm had been badly fractured and when medical staff at a hospital in Sana'a tried to fix with external bone fixation his arm got so badly infected that Muhammad went into a septic shock. Through the intervention of one of Sa'dah's leading traders, and Muhammad Abdullah Salih, the president's brother, who twisted the arm of Air France, Muhammad was delivered to a Paris hospital in a coma. However, after amputation of his arm he improved and already 10 days later the nursing notes said something to the effect that his relatives were behaving a bit obnoxiously by trying to smuggle special food [not *qat* though] into the ward.

5 Social and Political Conditions and Water Use

Introduction

This chapter argues that crop choice and water use patterns are not driven by purely economic considerations alone; rather, they are strongly influenced by tribal-political notions of power, knowledge and interests and by perceived and/or real socio-political and socio-economic values. Both over-abstraction and conservation of its groundwater resources can be understood in terms of the ability of actors to control and/or resist other actors. Moreover, perceptions and beliefs often explain why options which would appear economically sound and ecologically sustainable are subordinate to notions of tribal autonomy and food self-sufficiency.

Figure 5.1 captures some of the factors and values that explain water use patterns in the Sa'dah basin. First, as indicated by the box to the far left, labelled 'Socio-Political', political and economic instability, caused and exacerbated by the Gulf crisis, the 1994 civil war, the dispute with Eritrea and by the unresolved Saudi-Yemeni border issue, have fostered a climate of fear and uncertainty. Importantly, these perceptions have reinforced tribal notions of autonomy and food self-sufficiency. This is explained, in part, by the persistent importance of cereal production in the area. Secondly, the next text box to the right, 'Socio-Economic', refers to the demand for livestock in Saudi Arabia which drives livestock production in the Sa'dah basin with negative consequences for the area's groundwater balance. Moreover, socio-economic values of livestock explain the scale of sorghum and alfalfa production which, due to the loss of much grazing land, needs to be raised on groundwater irrigated fields. Crucially, the demand for alfalfa determines irrigation of cash crops and accounts for the lack of productive and allocative efficiency. Thirdly, the label 'National Policy' signifies government incentives and government policies that are responsible for groundwater over-abstraction. In this context, the production of cash crops, fruit and qat, in particular, will be investigated. Finally, to the far right, 'Environment' denotes that high levels of salinity cause many farmers to opt for flood and basin irrigation methods in an attempt to leach the soil. Throughout this chapter it will become evident 'that values are based on perceptions – and perceptions are what make and 'unmake' water a security issue' (Allan, J.A., 1998, personal communication).

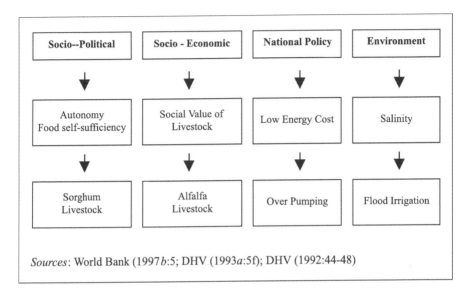

Sources: World Bank (1997*b*:5; DHV (1993*a*:5f); DHV (1992:44-48)

Figure 5.1   Water using factors in the Sa'dah basin

Livestock: The average number of livestock ranges from 10–24 head per farm.

Sorghum: 85 percent of farmers grow sorghum on up to $^1/_3$ of their land.

Energy cost: Diesel fuel, needed to run the water pumps, is priced at $^1/_4$ of its equivalent international level.

Salinity: EC (electrical conductivity) values: a mean of 953 microSimens/cm, between 1500–4100 mS/cm in eastern parts of the Sa'dah basin.

Livestock Production

> *ahamm shay – al-ghanam*
> Sheep are the most important [part of the farm]
> (Sa'dah farmer)

Figure 5.2 indicates the different values associated with the production of livestock and guides the following discussion.

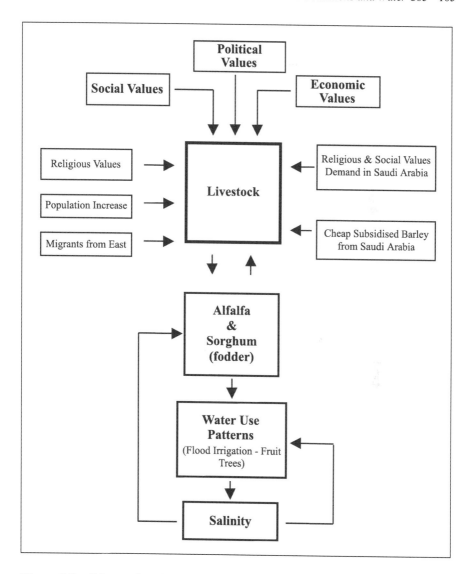

Figure 5.2   Livestock production: impact and consequences

Social Values

Births and festivals Livestock are important for social events and religious occasions. At the birth of a child at least one sheep but preferably two are slaughtered to celebrate the occasion. For the first ten days following the delivery the mother is given a lot of meat to eat, a tradition observed in many other parts of Yemen (Yemen Times October 4:1998). Also, during the first forty days after the birth, during which the mother enjoys special privileges, her female neighbours will honour her with gifts of meat.

Other occasions require sheep to be slaughtered. The ritual sacrifice of a sheep at the Muslim festival of *id al-adha* is a religious obligation for all Muslims. Usually, every family slaughters at least one sheep. Only one third of the meat should be kept, the rest must be distributed to the poor. But frequently, those with more means slaughter more than one animal. Beside *id al-adha,* sheep are slaughtered on two other Muslim festivals, on the day marking the birthday of the Prophet Muhammad (*mawlid al-nabi*) and at the end of Ramadhan (*id al-fitr*).

Even outside the major Islamic holidays, the cultural and religious values associated with the possession of livestock are significant. It is common practice to slaughter sheep in gratitude to God after an individual has been healed from a disease or after the successful recovery following an operation. Having been part of a medical team at Sa'dah's Republican Hospital for several years, the author has witnessed on countless occasions the slaughtering of sheep and goats in response to a perceived blessing, usually healing or medical treatment.

In addition, the slaughter of beasts is symbolically significant in tribal conflict mediation (Adra 1985:281). In the case between sections of the Hada and Bani Mu'adh tribes, referred to in the previous chapter, tribal peace was finally restored only after no less than fifty sheep and goats had been killed. In another case, where the death of a young girl, following a routine operation, led to an official investigation, reconciliation between the foreign hospital staff and the relevant tribe was eventually achieved through tribal customary procedures (*urf*), which required the slaughter of three sheep. Attempts to secure the release of a colleague kidnapped by a tribe to the east of Sa'dah in January 1999, further demonstrates the cultural and tribal-political value of livestock. After initial government efforts to negotiate with the tribe had proven unsuccessful the leading tribal dignitaries from the Sa'dah basin planned to intervene by taking a large number of sheep for slaughtering in the mountain territory of the kidnappers. Subsequent developments rendered these tribal-customary procedures unnecessary but the plan indicates the importance of livestock in Sa'dah's tribal society.

Migrants from the east   A prolonged period of drought between 1980 and 1987 that affected, in particular, the Wa'ilah areas to the east of the Sa'dah basin forced many of these semi-nomadic herders to migrate into the basin. One family, for example, came with 200 sheep and goats when they first settled near the Sa'dah airport in

1981. In the absence of grazing areas the daily demands for fodder placed a heavy burden on the family and they have since reduced their herd to about 80 head. But even such numbers require large amounts of barley as well as sorghum fodder. This they raise not on the new farm but on wadi tracts in the eastern Wa'ilah areas where the family owns land which is irrigated by spate (*sayl*) irrigation.

In conversation with Bedouin who migrated from the eastern arid region to the Sa'dah basin it became apparent that their livestock numbers have gradually dwindled, from a few hundred to an average of fifty head per family. In contrast, those in the business of buying and selling sheep have observed the opposite trend among the settled Sahar tribes of the basin. One Bedouin with fifty years of trading in livestock expressed it as follows; 'the people of *Sa'id* (Sa'dah basin) now all have more sheep than they used to have. Before the 1980s lack of water and fodder limited the size of flocks, now farmers have plenty of water and are able to grow alfalfa on their farms.'

The same development has also been observed in the context of the central highlands where Dresch (1989:345) states that the number of livestock worth keeping used to be limited by the long journey to the waterholes. But over the last two decades diesel pumps and wells have lead to an enormous increase of livestock. He quotes a man from Sufyan with land in Arhab who increased his flock from 50 to 400. Consequently, grazing rights became a major issue of contention.

Population increase and livestock numbers   Through a combination of on-farm observations and interviews it can be established that total numbers of livestock in the Sa'dah basin have increased over the past two decades. The majority of farmers visited kept between 15–20 sheep and goats. In the area of Hamazat, where farmers have access to grazing, average flock sizes were even higher, 20–30. In the Buqalat area, just east of the old town of Sa'dah, farmers have 20–30 head of sheep. This latter area also supplies alfalfa for the fodder demands within the town of Sa'dah where the average family keeps up to four sheep and goats beside or above the dwelling. However, in the new section of town where people from al-Mahadhir have settled, livestock numbers are comparable to those on the rural farms. And even in the arid and rocky terrain inhabited by Al Ammar, south of the Sa'dah basin, families keep an average of 15 sheep per household. The case of a farming family from the Bani Mu'adh is indicative of the overall trend. The now ageing head of a family recalls that before the 1962 revolution, when he got married, his father owned 40 sheep and 20 goats. Now, in the 1990s he and his two brothers each keep 20, 25 and 30 sheep and goats. Considering the tremendous increase of population due to migration and natural increase (**Figure 5.3 and 5.4**) overall livestock numbers have risen significantly.

What is significant in this context is the fact that in contrast to the earlier pre-groundwater exploration period, fodder demands for livestock now have to be met by groundwater irrigation. Privatisation of tribal common lands and the rapid spread of groundwater irrigated farms has resulted in the loss of large grazing areas.

Population Figures

The figure below shows the large increase of population in the Sa'dah basin over the last 20 years. The region's high child mortality rate has fallen rapidly over the same period as medical services have become available. Whereas women aged 40 and over have usually gone through more than 10 pregnancies with only 2-3 children alive today, women aged 25 –30 have had about 10 pregnancies with 7-10 children alive. Interesting, too, is the fact that town girls – considered educated – are married off earlier than farm girls. This is to protect them from the perceived moral dangers of the city whereas rural girls are given in marriage at a later age because their help is needed on the farm.

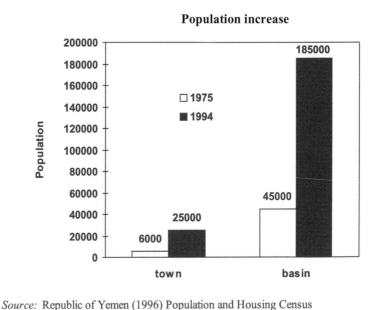

Source: Republic of Yemen (1996) Population and Housing Census

Figure 5.3   Increase of population in Sa'dah basin between 1975 and 1994

Figure 5.4   Population growth
A group of young boys getting a ride on the back of a Hilux pickup from school in Sa'dah town. Yemen's population grows by 3.7 % each year (Brunner 1999:130).

Whereas in other regions of Yemen the education of girls (traditionally herding the sheep) may have resulted in the reduction of livestock numbers per household (Kopp 1984:80), this is not the case in the Sa'dah basin where alfalfa and sorghum fodder can be produced by well irrigation.

From the above it is clear that livestock are viewed as an essential and important part of society in the Sa'dah area. The experience of one informant exemplifies best the enormous social and cultural values of livestock. Muhammad is from a small rural community of the Bani Mu'adh.

For a number of years he lived in Sa'dah town, where he worked for the Ministry of Public Health. When he could no longer afford the cost of renting a flat for his family he decided to move back to his village. From there he now commutes daily the short distance to his work in town. But no sooner had he moved back than villagers started to ridicule him for lacking the essentials to prove himself part of the community. 'Buy yourself some *ghanam* (sheep), without livestock you are neither *gabili* (tribesman) nor *muzara* (farmer). What are you going to do when you receive guests, celebrate the *id* (Islamic festival) or the birth of your offspring?'

Economic Values

Apart from the perceived social values attached to livestock Sa'dah's politicised environment provides strong economic incentives for the production of livestock in the area. The dynamics of this process are captured at the top-right corner of Figure 5.2.

Demand in Saudi Arabia for livestock is a strong influence on the production of livestock in the Sa'dah basin. The Muslim pilgrimage of *hajj* to Mecca is a religious obligation for every Muslim. About two million Muslim pilgrims perform the *hajj* every year, which has become Saudi Arabia's second-most important source of foreign income earnings, exceeded only by the country's revenues from oil. The religious duty to sacrifice a sheep at the *hajj* festival pilgrimage has already been mentioned. As a result, demand for livestock in Saudi Arabia peaks during this season. Farmers in the Sa'dah basin have a strong incentive to produce livestock for the Saudi market. The unregulated nature of cross-border trade makes it impossible to arrive at precise figures. However, through interviews with farmers and discussions with livestock herders and as a result of observations at tribal markets and along mountain tracks a number of conclusions can be reached.

Cross-border trade   Substantial cross-border trade in livestock is conducted for several months every year. This starts before Ramadhan and lasts till the *id al-adha*, the 10th day of the *hajj* when the sacrifice is performed. In 1996, a few weeks before the hajj, it was observed that most four-wheel drive pick-ups, travelling along tracks towards Saudi Arabia, were loaded to capacity, with up to 25 sheep. Many farmers sell directly to middle men but loading points for livestock also include two or three weekly tribal markets in the Sa'dah area. During this peak-season an average of ten carloads (10–25 heads per car) are reportedly taken from the large market of Suq al-Talh every week. Over a period of five months this would add-up to an average of 4,000 sheep and goats from the market of Suq al-Talh alone.

Tribal networks   Transferring sheep and goats untaxed to the Kingdom is made possible through social links and tribal-political networks and by the region's proximity to Saudi Arabia, which gives the Sa'dah basin an important comparative advantage. In addition to a number of main routes, which serve as approaches to the Saudi Kingdom, there are numerous other tracks from the Sa'dah basin which lead to Saudi Arabia. The routes used depend on tribal links and tribal-territorial considerations. One mountain track said to be used to transport livestock from the Sa'dah basin takes as little as two hours to reach the Saudi side. Compared with the risky business of smuggling consumer goods, prominent in the 1980s, 'trade' in livestock has obvious advantages. Sheep and goats can also be herded across over inaccessible mountain tracks.

Incentives and subsidies   The rearing of livestock in the Sa'dah basin is helped, in part, by cheap, Saudi-subsidised barley (*sha'ir*). At the weekly tribal market at Suq al-Talh a large section is allocated for the sale of local and imported cereals. Barley (50kg sacks) is bought in Saudi Arabia for SR 25 per sack (US $ 6.6) and sold at the Talh market for YR 1,400 (US $ 11–12, 1997 figures). The green barley sacks clearly reveal their origin (**Figure 5.5**). Most of this barley enters Yemen unofficially by *tahrib* (smuggling). Mountain tracks linking the Saudi province of Asir with the Yemeni area of Munabbih, to the north-west of the Sa'dah basin, are used for this type of cross-border transfer. However, on occasions and for limited periods of a few months only the Kingdom permits certain Yemeni entrepreneurs to import Saudi barley officially. Saudi barley is called *ahmar* (red) and considered inferior to the local (*baladi*) variety known as *abyad* (white). Consequently, it is used as fodder for sheep and goats only. In contrast, locally produced barley is regarded as 'priceless' – 'no one will sell *sha'ir baladi*', as barley bread is highly valued in combination with certain traditional dishes.

Demand   Demand in Saudi Arabia for Yemeni meat creates strong economic incentives and provides income in hard Saudi currency. The 1990s cross-border trade promises secure and stable incomes, in a period of high inflation, a stagnant economy and highly fluctuating prices of local markets. A large sheep will fetch Saudi Riyal 600 (US$ 160 or YR 15–20,000, 1996 figures) across the border compared with only YR 6,000–9,000 on the local market. The high and 'secure' returns from the rearing of livestock have led some farmers to shift away from the uncertain profits associated with many cash crops due to a lack of market control mechanisms. Moreover, demand in Saudi Arabia has lead to an increase in the price for meat at local markets. This explains why many in the town of Sa'dah keep a number of sheep in their house. The cost of having to purchase a sheep for social and religious occasions would consume a month's salary of most civil servants.

Livestock and Sorghum

The box below 'Livestock', labelled 'Alfalfa and Sorghum', as seen in Figure 5.2, indicates that livestock raising drives the production of sorghum and alfalfa. Local varieties of sorghum are highly valued and have their special place in the traditional cuisine of the Sa'dah area. However, this only partially accounts for the fact that farmers allocate, on average, up to one fourth of their irrigated farm area to the production of sorghum (DHV 1993:5). Farmers in the study area agreed that the high fodder volume of sorghum is a primary reason for growing groundwater-irrigated sorghum (**Figure 5.6**). Kopp (1981: 151 n.167) points out that the fodder value of *dhurah* (sorghum) is, in most cases, even above that of the grain. In the

Figure 5.5   Barley from Saudi Arabia
Barley is sold mainly as fodder for raising livestock at the tribal
market of Suq al-Talh. The sack indicates the origin as Saudi Arabia.

Figure 5.6   **Carrying fodder from the sorghum harvest**
The fodder value of sorghum has often been overlooked in agro-economic cost-benefits analysis done by economists unfamiliar with the socio-economic aspects of a given culture.

study area sorghum grows two to three metres tall. Its stem (*qasab* or *sund*), leaf (*sharyaf*) and head (*hamad*) provide essential inputs for the rearing of livestock. The sorghum leafs (*sharyaf*) are picked by the women a few weeks before harvesting. After harvesting the empty sorghum head (*hamad*) is highly valued for its nutrients. For the evening feeding time *hamad* is boiled and mixed with dry bread and some sorghum grain. It is thought to increase the animal's milk production. Sheep prefer the fodder (*alaf*) of white sorghum (*dhurah bayda*) to red sorghum (*dhurah hamra*), the grains of which are preferred for human consumption. To keep a herd of twenty sheep and goats a farmer will crop about 100 *hablah* (0.25 hectare) of sorghum and half of that for alfalfa. However, on the average two and a half hectare farm, the total area allocated to the production of sorghum is likely to be twice that amount (0.5 hectare) as it also reflects the high value of sorghum within the area's traditional cuisine.

Livestock and Alfalfa

The production of alfalfa is also driven by the values attached to sheep and goats. The Sa'dah basin has always been an important centre for the rearing of livestock (Gingrich and Heiss 1986:27; 106). Up to the mid-1970s there was ample grazing and therefore limited demand for alfalfa. Privatisation of tribal common lands from the mid-1970s onwards and the consequent extensification of groundwater irrigated agriculture changed the traditional way of maintaining flocks. The loss of grazing has meant that demands for groundwater-irrigated alfalfa production has increased rapidly. At the same time, access to and availability of groundwater has encouraged farmers to keep sizeable herds even after loosing grazing areas. Allocation and irrigation patterns reflect the demand for alfalfa. This is indicated in Figure 5.2 by the large text box 'Water Use Patterns' and will be explained later in this section. The area allocated by individual farmers to grow alfalfa depends on the number of sheep they keep. Between two and four *hablah* (25–100 square metres) per sheep and 30–50 *hablah* per cow of alfalfa is considered necessary for one livestock unit. Most farming households with access to additional grazing will irrigate about forty *hablah* (0.1 hectares) of alfalfa, which reflects an average flock size of twenty sheep and goats.

Livestock in Sa'dah Town

The situation in the vicinity of Sa'dah town is different from the farms further away. Here, alfalfa is grown on many farms as a cash crop in response to the rising demand by livestock kept in the town itself. As shown earlier in Figure 5.3 the Sa'dah town population increased from 6,000 to 25,000 between 1975 and 1995. A minimum of two sheep are fattened in most homes. Where space is at a premium livestock are kept on the top of the flat roofs. For all the reasons mentioned earlier livestock are considered essential. Demand in Saudi Arabia has pushed meat prices up and many inhabitants in Sa'dah town would not be able to afford to buy a sheep in order to satisfy religious and/or social obligations. Besides, rearing a few sheep is considered a smart investment, comparable to a 'savings account', and a source of cash which can be tapped when required (DHV 1990:19). The sale of a six months old lamb for YR 6.000 (US $ 48, 1997 figures) adds a considerable amount to the budget of a civil servant with a salary of only YR 8,000 (US $ 64 per month).

A survey in one section of the town revealed that livestock numbers per household were between five and seven head. In a new and more spacious section of Sa'dah town migrant families from the area of Mahadhir keep between twenty and thirty head each. Relying on alfalfa from the market they are prepared to spend YR 150 and more per day (half the wage of a day labourer) to buy their daily supplies of alfalfa.

The total number of livestock in the town of Sa'dah (pop 25,000) is estimated at over 6,000. This figure is based on assuming that Sa'dah's 3,738 households (Republic of Yemen 1996:309) keep a minimum of two sheep per family unit only. However, actual livestock numbers must be much higher. As mentioned above, even families in the crowded quarters of Sa'dah's old town often keep up to five sheep. And in a number of newly developed areas of Sa'dah town livestock numbers exceed ten per household.

This figure then could translate into the demand for at least 6,000 *hablah* of alfalfa (15 hectares) around Sa'dah town. Given the high crop water requirements (CWR) of alfalfa, 17,970 cubic metres/hectare (DHV 1993:38), the total water demand would be 269,550 cubic metres. This represents thirty litres of water per day for every inhabitant in the town, more than the current daily consumption per head in most families. This figure becomes more significant when considering that present water supplies are unreliable and insufficient in most parts of the town while non-existent in many others.

Alfalfa from farms nearby is sold at the Sa'dah market (**Figure 5.7**). The early morning sees Yemeni males pushing wheelbarrows to market in order to buy the quantities of alfalfa needed for the day. There are also other outlets in the town's various living quarters. Selling alfalfa is usually the task of women and its marketing provides them with a chance for interaction and some personal income. For the capital, San'a, Mackintosh-Smith (1997:12) mentions the huge piles of alfalfa sold by women in front of their homes. His remarks indicate that many city dwellers in San'a keep livestock in their homes.

Until a few years ago sheep and goats were left to roam Sa'dah's streets and alleyways unattended to search for garbage and eatable stuff. A few were stolen or disappeared and now town people no longer let their small flocks wander. Alfalfa, which is grown in the vicinity of the town, and imported wheat, which is available at subsidised prices constitute the primary sources for fodder for the town's high livestock numbers.

Salinity

Salinity provides one answer why alfalfa for the Sa'dah market is grown on farms to the west, east and south of the town. During the 1982/83 well inventory, electrical conductivity (EC-values) here measured already in excess of 1000 mS/cm compared to values below 750 mS/cm for most of the Sa'dah basin.[1] By 1992 critically high values EC values were measured in these areas, between 1500 and 4100 mS/cm. In serveral wells the level of salinity had more than doubled over a period of just nine years (DHV 1992:48).

In the Buqalat area, just east of the old town of Sa'dah, this problem might have become more pronounced because of the decline of groundwater levels over the past

**Figure 5.7   Alfalfa and other animal fodder being sold at the Sa'dah town
market**
Cutting and selling of alfalfa is usually the task of women as seen here
by the black-veiled women at the right of the photo.

decade, from 40 to 80 metres (DHV 1992:44), although the relationship between
increasing depth to groundwater and increasing levels of salinity can not easily be
established and there may be other hydrogeological factors responsible for the
salinity problem to the south and east of Sa'dah town (DHV 1992:48). Whatever the
reason, in the Buqalat area high salinity leaves farmers few options apart from
growing alfalfa (**Figure 5.8**), which tolerates saline water and improves soil
conditions (Kopp 1981:98). This is indicated in Figure 5.2 by the arrow that links
the text box 'Salinity', at the bottom, to the text box 'Alfalfa and Sorghum'. In the
Buqalat area problems associated with groundwater scarcity and salinity have forced
some farmers to abandon farming. Others have shifted increasingly to alfalfa, the
production of which has become a coping strategy and a way for survival, be it
short-term. First, alfalfa can be cut every two weeks for a period of up to four years.
Secondly, yields from a small 0.1 hectare plot, sold at the near-by Sa'dah market,
provide a farmer with about YR 9,000 (US$ 72) per month, which is more than the
monthly salary of the average civil servant. Thirdly, as many families have come to
struggle for survival, the daily returns of YR 300 from alfalfa provide urgently

Figure 5.8   A farmer prepares his fields for the cultivation of alfalfa

needed cash. And lastly, since the production and harvesting is predominantly the responsibility of women, it leaves men free to seek additional income through work in Sa'dah town.

Livestock and Irrigation Patterns

Demands for feed to raise livestock drive irrigation activity as indicated in Figure 5.2. Prompted by government policy and enticed by economic and political incentives (Ward 1997:5) most farmers in the Sa'dah basin shifted to fruit production in the mid-1980s, especially citrus (Lichtenthäler 1999). Flood irrigation has been applied for a number of reasons. First, in the mid-1980s groundwater appeared to be available in abundance and pumps were operated at low cost. Secondly, compared to pomegranates and grapes, which do not require water for 3–4 months after their harvest in early autumn, citrus trees, like alfalfa, need to be irrigated regularly over the whole year. Especially during their initial three to five year maturing period and while the trees were still small flood irrigation was applied in the orchards to grow alfalfa between the rows of trees. Thirdly, farmers rightly argued that flood irrigation would decrease the salinity in the soil and increase its productivity (Kopp 1981:98). Consequently, the root system of the trees has developed in response to irrigation method, frequency and supply. Some traders have experimented with drip-irrigation, perhaps in the hope of marketing the idea and trade in modern irrigation equipment. But all those who tried abandoned its use claiming that the mature trees soon suffer from drought stress. Problems associated with salinity are likely to be exacerbated by the on-going mining of Sa'dah's aquifer (DHV 1992:48). Even today, farmers planting new trees continue to opt for flood and basin irrigation practices in order to counteract the negative effects of salinity. As shown in the link between 'Salinity' and 'Water Use Patterns' at the bottom of Figure 5.2, this process completes the [vicious] cycle of excessive groundwater use in the area.

The above section has analysed the various values of livestock in the Sa'dah basin. It was shown that livestock production determines crop choices and water use patterns. Demand for fodder explains, from an economical perspective, the lack of productive and allocative water use efficiency methods in the area. Almost all actors in the Sa'dah region have an incentive to raise livestock in a politically distorted rural economy in which they find themselves. Cultural and social norms and short-term economic signals reinforce the raising of sheep. The long-term value and role of water is discounted. The water resources of the region are being diminished and degraded. At the same time Sa'dah's politicised environment fosters a political economy in which skilful actors can thrive by evading taxes as well as capitalising on the various subsidies on offer from both governments.

Qat

> Telling farmers that they should not grow qat, however, is like trying to convince the National Rifle Association that Americans do not need guns (Varisco 1983:33).

Introduction

Qat is another crop which is of high cultural and social significance. Qat has an important socio-economic niche in the Sa'dah political economy. Unlike livestock rearing qat is relatively benign in terms of the impact on the region's groundwater.

This section will analyse the various values associated with qat. Qat is the most important cash crop in Yemen and in the study area. It will be shown that Sa'dah's politicised environment creates strong economic incentives for its production. This becomes evident by examining the lucrative but risky cross-border smuggling into Saudi Arabia, where the possession and consumption of qat is a criminal offence and punishable by imprisonment, according to Sa'dah qat traders. In the same context, the ability of local actors to evade the control of the Government of Yemen and its tax requirements will be highlighted. In addition, it will be argued that qat meets many criteria with regards to water use efficiency (productive and allocative).

No foreign visitor to Yemen can avoid puzzlement at the daily sight of Yemeni men carrying what appears to be bunches of flowers wrapped in fine cellophane. 'These people really love their spouses' thought the writer at his first visit to Yemen in 1984, in complete ignorance of the almost institutionalised daily ritual of qat chewing. The fresh sprouts of the evergreen *Catha edulis* bush are chewed for their stimulant and mild narcotic effect. Although known as early as the 14th century, the chewing of qat has become widespread throughout the country since the revolution in 1962 (Weir 1980:71,85). While the benefits of qat are extolled by many Yemenis, western visitors to Yemen usually 'present the whole practice in a highly unfavourable light' (Weir 1980:54).

Qat is a politically sensitive issue as well. Attempts by Prime Minister Muhsin al-Aini in 1971–72 to stop the habit, which he believed would ruin Yemen's society and economy not only failed but eventually caused his downfall (Weir 1980:67).

Socially, qat is of immense importance. Qat often affects the main eating habits. The timing of daily activities is structured by its time of consumption. It is an important part of all markets and 'no other society is so influenced by the use of a drug as is North Yemen' (Kennedy 1987:101). The failed coup d'etat in 1948, as well as the successful one in 1962, are said to have been planned out during daily qat sessions (Schopen 1978:138).

History of Qat

Evidence from historical texts is inconclusive but suggests that qat was chewed as early as the fourteenth century in the southern region of the Red Sea, mainly Yemen and Ethiopia (Weir 1980:71).

At first, its use seemed to have been restricted to the Sufi (mystical) fraternities and the religious aristocracy of the *sayyid*s (pl. *sadah)*. Especially the Sufis used the drug to enhance their experience of religious ecstasy in reaching a mystical union with the Divine. In Sufi poems, qat became metaphorically described as 'wine of the believers' and the 'stairway to paradise' (Schopen 1978:52).

When in the 17th century Yemen's coffee export trade reached its zenith, the use of the stimulant spread to merchants and traders. Only after the decline of the coffee export in the beginning of the 19th century did the cultivation of qat increase considerably and people at all levels of society began to chew. A variety of different qat qualities soon emerged to maintain the differentiation of social status. While the aristocracy (*sayyid*s) continued to use qat for religious purposes, the majority of the population chewed the leaf to help them in their hard physical labour (Schopen 1978:52).

The almost institutionalised use of qat throughout the whole country became possible only after most villages became accessible by some kind of road or track during the 1970s.

Geography of Qat

Qat grows successfully in altitudes of 1200–2400 m with annual average temperatures of 16–22 C. and between 400–2000 mm precipitation. Therefore, most of the qat is grown in the western mountain areas. The number of terraces (49 per cent of cultivated land in Yemen is on terraced plots) on top of each other can amount up to several hundred. 250 terraces have been counted by Rathjens within the span of 1000 metres altitude. Fields are often very small and not easily accessible. They can be a distance of 3–4 km away from the farmer's house and the farmer might have to descend more than 800 m to cultivate them. Farming at these conditions is extremely labour intensive. But whereas the cultivation of grain is very labour intensive and requires the ploughing of the soil up to 15 times until the harvest, qat cultivation has a number of advantages, especially when one considers the labour shortage in the 1970–80s due to migration to Saudi Arabia and the low price of imported wheat during the same period.

The Sa'dah basin does not have a long tradition of growing qat. Until the 1960s no significant amounts of the shrub were grown in the area. Traditional qat producing areas are further west, the mountains of Razih in particular. The Wa'ilah to the east of the basin have never much taken to qat and even in the early 1980s only a negligible number of qat was reported in these areas (Gingrich and Heiss 1986:120). Until the Revolution in 1962 chewing qat was an expression of social status *ibarah an mustawa ijtima'i* and, at the time, associated with the upper strata of society *al-tabaqah al-ulya* (Informants from Sa'dah basin). What little qat was grown supplied the Imam's resident emir (Chargé d'affaires) in Sa'dah and his aides, the religious scholars (*sayyid*s).

Qat trees can produce for over fifty years (Kopp 1981:236). In the area of Al Ammar, now famed for its quality qat, individual trees that supplied the emir and his entourage during visits to Wadi Sharamat, can still be pointed out, towering high in the midst of mature grape orchards. And like the rainfed vine around them these trees have the ability to endure water stress for prolonged periods, up to six months.

Economic Values

Figure 5.9 illustrates the various values of qat, as described and perceived by the local Sa'dah farmers. It also shows the relationship between qat production and water use efficiency. In the following discussion it will be shown that a) Sa'dah's politicised environment adds extra value to qat, and b) that qat has qualities which make it a highly suitable crop to adjust to increasing water and water-related stress such as is experienced now by many farmers in the Sa'dah basin.

Ideal for poor soils   Qat is ideally suited for poorer soils. This was pointed out by Swanson (1979:36) who reports that a village located in a dry zone used the little water available to irrigate qat fields on some marginal poor land. While qat provided

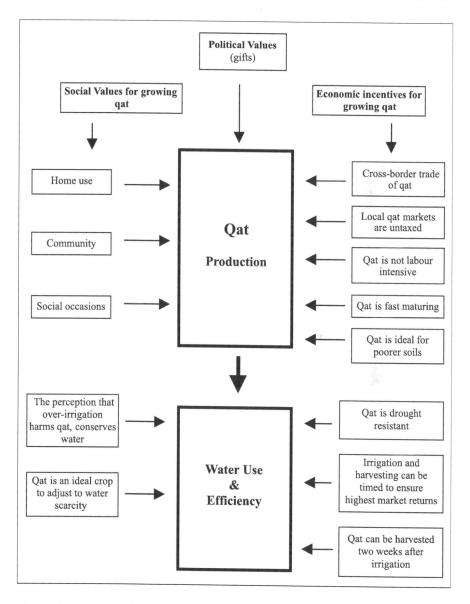

Figure 5.9   Qat: values, incentives and water use

a relatively profitable crop on the poor land, the land would otherwise have produced only a poor grain crop. Similar perceptions affect qat production in the Sa'dah basin where it has been observed that qat tends to be grown on poorer and sandier soils. This is, in particular, the case on the basin's western fringes where floods and runoff from the steep escarpment have caused soil erosion on these tracts. Given the basin's incline of 0.5–1 percent (TNO-DGV 1985:6) from west to east, runoff has washed the better soils towards the central areas of the basin, making these tracts more suitable for the production of citrus. By contrast, in many of the slightly more up-stream areas of the Bani Mu'adh soils tend to be poorer and sandier making them ideal for qat cultivation, according to many local farmers. In these western tracts over 50 percent of irrigated farmland is allocated to the production of qat.[2] This compares with an earlier study in 1992, which found that 'many farmers devote up to 30 percent of their farm holdings to it' (DHV 1993a:5).

Fast maturing   Unlike citrus, which starts to produce fruits only after four years initial returns from qat can be expected after the first year. Thereafter, the qat tree produces a harvest every four months if irrigated (*masna*) or every six months if the farmer relies on water from rain and runoff (*anthari*). Returns from qat vary according to season and supply. In the 1980s the returns from qat production were phenomenal but due to increasing supply net returns have decreased since. For the Razih mountains to the west of the Sa'dah basin Weir (1985:72) figures market values of YR 400,000 (US$90,000) for one hectare of land planted with qat trees. This compares with DM 330,000 (US$200,000) per hectare per year from qat produced near the main qat markets of the capital San'a during the same period (Betzler 1984:67). In 1996/97 a farmer in the Sa'dah area could expect between YR 100,000 and 150,000 per year from a small area of 50 *hablah* (0.125 hectares). This amounts to US$ 8,000 (1997 figures) per hectare per year and compares badly with Betzler's figures of approximately US$ 200,000 per hectare per year. But the enormous difference is explained by the devaluation of the Yemeni Riyal since the mid-1980s. In 1984 US$ 200,000 had an exchange rate value of one million Yemeni Riyal (5 Riyal to US$). In 1997 US$ 8,000 was exchanged for the same amount, one million US$ (125 Riyal to US$). Hectare values from qat in Yemeni Riyal between the early 1980s and the late 1990s are similar. In real terms, however, profits from qat have also fallen considerably over the past decade.

In the Bani Mu'adh area many farmers irrigate up to one hectare of qat with a present market value of YR one million per year. Compared with the average wage of a civil servant in Sa'dah town, who can expect no more than YR100,000 per year the incentives for qat production continue to be high.

Not labour intensive   Qat is a crop which requires a minimal amount of labour (Kopp 1977:32; Lichtenthäler 1996:123). Especially, during the late 1970s and early 1980s when, according to local sources, one half of Sa'dah's adult population was involved in cross-border trade and/or worked in Saudi Arabia qat farming was ideal as it required very little labour. Moreover, agricultural inputs are minimal. Qat does not require any fertilizer. In fact Yemenis claim that fertilizer, if used, spoils the taste and lowers the quality of qat. Qat trees are usually dusted with *turab* (soil dust) twice a year (compared to 15 times in the case of grapes) to keep insects off. Commercial pesticides and growth hormones have only recently been applied but there is growing concern that plants thus treated may be harmful to health.

Compared with qat cereal production is much more labour intensive. At the same time returns from qat have been reported as 24 times higher than sorghum and even 3.5 times higher than coffee (Kopp 1981:238). Before the mid-1980s, when the government fruit import ban created economic incentives for cash crops, especially fruit, the increase of qat cultivation was often at the expense of grain production (Weir 1985:75). With cheap subsidised wheat and wheat flour flooding the Yemeni markets and as most regions suffered from labour shortage as a result of outward migration there was little incentive to produce wheat. In economic terms qat was a sensible and rational choice.

Local and urban markets   Between 1975 and 1995 the population in the Sa'dah basin has grown from 45,000 to 185,000. During the same time the town of Sa'dah has experienced an increase from 6,000 to 24,000 inhabitants (Republic of Yemen 1996).[3] These figures also reflect the increase in local demand for qat. Quist (1990:63) suggests that as many as 60 percent of Yemeni men and 30 percent of Yemeni women participate in the qat session most afternoons spending half of the average monthly income on qat.

The quality of qat depends not only on its place of origin but on its freshness. If at all possible qat is picked in the morning and must be sold by noon time the same day. In spite of improvements in the region's infrastructure over the past decade traditional qat-growing areas can be several hours way. From the mountains of Razih, for example, it can still take between 4–6 hours to reach the Sa'dah basin. This gives Sa'dah's qat farmers considerable advantage over suppliers from the western mountains. Moreover, sudden and heavy rainfall can inundate mountain tracks and interrupt qat supplies from regions outside the Sa'dah basin. As a result local qat dominates the markets during these times. During and after outbursts of rain they have to do emergency repairs to protect the roof and the walls of their mud houses. Their grapes, which are usually spread out on the flat roofs of the multi-storied tower houses to dry must also be rescued and protected. Initially, after spells of rainfall, the price for qat goes up as supplies come only from nearby farms.

Sa'dah's comparative advantage is especially noticed during the dry season. Qat farmers with access to groundwater and nearby markets have an advantage over those in the rainfed areas since they can time irrigation, cutting and marketing of qat to coincide with the dry season when little qat from the traditional qat growing areas is available and when, as a result, prices are higher (Weir 1985:87). At any time during the year fresh qat leaves can be harvested as soon as too weeks after the trees have received sufficient water. This is very attractive to farmers as it allows them to control harvesting. Equally important, returns from qat can be spread out over many months and provide cash income on short notice during emergencies. This is especially attractive for poorer farmers suffering from diminishing supplies of groundwater, for whom qat acts as a 'safety net' and a means to survive. Therefore, with respect to timing the harvest and taking advantage of favourable market conditions, qat is endowed with unique qualities, which no other locally produced cash-crops can match (Tutwiler and Carapico 1981:52).

Qat and cross-border trade   The ability of the Sa'dah region to respond to demand for qat in parts of the Saudi Kingdom, where it is outlawed, provides strong economic incentives and adds extra value to qat production.[4]

Until the Gulf Crisis in 1990/1991 large daily qat markets in the Sa'dah basin existed only in the town of Sa'dah and in Suq al-Talh. Outside the Sa'dah basin Suq Al Ammar, just 35 km south of the provincial capital and along the main road to San'a, is renowned for the high quality *Ammari* qat produced in this rocky and arid terrain.

In the 1990s, and especially after the Gulf Crisis, qat production in the Sa'dah basin has increased. The setting up of new qat markets during this period reflects the ability of tribal actors and groups to resist the control of other actors and, especially the government. And, as we shall see in the next chapter for many qat production has become a necessity in order to survive.

Suq al-Khafji is a large wholesale market for qat, located about 12 kilometres north of Sa'dah town along the main road leading towards Saudi Arabia. It was started shortly after the re-capture the town of Khafji[5] during the 1990/1991 Gulf Crisis.

By that time the largest tribal market in the area, Suq al-Talh, could no longer cope with the chaos created by the increase in qat trade. As a result qat traders moved out of Talh and set up Suq al-Khafji. This shift has also meant a shift of control, away from Talh, which is dominated by politically and economically powerful tribal figures, to Walad Mas'ud (Sahar), whose tribal territory, along with that of the Bani Mu'adh to the west make up the biggest part of the Sa'dah basin.

As a wholesale market for qat and in contrast to other qat markets, which boom with life around noon time, Suq al-Khafji is the busiest in the early morning hours. Qat from Suq al-Khafji is supplied not only from farms in the Sa'dah basin. Improvements in infrastructure over the past decade have meant that it can be brought in fresh from the surrounding regions, which, due to more favourable climatic conditions, have gained a reputation for high-quality qat. Especially during the rainy season car loads of qat arrive at Suq al-Khafji from Khawlan, 2–4 hours to the west of the basin and as far as Razih, an area 4–6 hour drive to the west with a history of growing qat (Weir 1980:86; 1985:68). From Suq al-Khafji an efficient network of traders ensures that qat is delivered to other markets.

Since the end of the 1991 Gulf war, which led to the expulsion of over one million Yemenis from Saudi Arabia, and consequently ended the large flow of foreign remittances, the demand for qat in Saudi Arabia has greater significance.

In Saudi Arabia possession and consumption of qat is forbidden and persons contravening the law can expect a hefty punishment (Morris 1986:123) of many years of imprisonment, according to informants. But in spite of these deterrents qat from the Sa'dah region is regularly smuggled across into the Saudi Kingdom. Over the past six years, in particular, cross-border trade in qat has increased as net profits from other agricultural products have fallen. The volumes of qat smuggled are substantial enough to keep the price of qat high. Locals involved in cross border activities reported that the price for qat in the Sa'dah basin drops significantly during times when Saudi border controls are tightened or when the border areas get sealed off, as happened over a period of four months during 1997.

Since chewing qat became a central part of Yemeni life and culture in the late 1970s qat has always 'leaked' into Yemen's powerful neighbour kingdom, notably from the Razih region. Due to the high risks involved, qat for the Saudi Market used to be dried and crushed into powder, which made it easier to hide and also allowed for longer delivery times since freshness no longer was a criterion. Now, most qat consignments for consumers across the border are said to be delivered fresh. As the quality and price of qat depends as much on its freshness as on the region of its origin Yemenis have developed ways to preserve the moisture of the qat shoots over a number of days. For this purpose qat is commonly packed inside the stem of banana trees. This use has helped to preserve banana production in the mountainous Razih area, according to Weir (1995:68,69).

Those involved or informed about the cross border qat trade figure that 4–5 car loads from the Sa'dah basin are taken daily to the Saudi border. Their Yemeni market value is considerable, between YR 700,000 and YR 1.2 million (US$ 5,600 – 9,600 in 1997 figures) depending on the season.

Points of entrée for qat are the areas north and north-west of the Sa'dah basin where Yemeni tribal groups share cultural values and social ties with a number of Yemen tribes now located on the Saudi side. Jabal Fayfa, annexed by Saudi Arabia since 1934, lies just across from the Yemeni mountain regions of Munabbih and Razih. Acording to informants many in Jabal Fayfa share the qat habits of their Yemeni neighbours across the border. Apparently Jabal Fayfa is the only area in Saudi Arabia where some qat is grown and where, for political reasons, the Saudi government has found it expedient to turn a blind eye to its production. But the limited amounts of qat tolerated in the Fayfa mountains may only be used for private consumption. Security checks at the border of Saudi Arabia Asir Province are aimed to prevent qat from reaching other areas.[6] However, Yemenis involved in cross-border qat trade report that qat from the Sa'dah region enters not only Jabal Fayfa but the other former Yemeni territories along the Saudi side of the border.

Suq al-Khafji is not the only place from which qat is taken to the Saudi border. Qat from the area of Al Ammar (*qat ammari*) is renowned for high quality and during visits to the Al Ammar's main market in October 1997 it was observed that a number of cars were being prepared to deliver qat to the border areas.

The area of Razih has, perhaps, the longest history of benefiting from cross-border demand. But during the dry season little rain-fed qat can be harvested on Razih's mountain terraces. This is the time when Razih's qat traders come to Suq al-Khafji to buy irrigated qat, which is then taken back to Razih from where it makes its way across the near-by border.

Alternatively, qat traders visit farms and negotiate directly a price for the purchase of an entire harvest. Qat is usually picked twice a year. This is reportedly the practice in much of Bani Mu'adh where up to 50 % of the irrigated land is now allocated to qat cultivation.

Qat markets   Since 1994 the Bani Mu'adh have established their own qat market. Suq al-Anad is located only five kilometre north of Sa'dah town within the tribal territory of the Bani Mu'adh. In contrast to Suq al-Khafji, further north along the same road, Suq al-Anad is busiest during the noon hours as it serves mainly local demand. Qat buyers even come from Sa'dah town to avoid paying the tax levied on the shrub by the government.

An incident that occurred at the end of the Civil War in 1994 shifted control over qat sales from Sa'dah town to what is now Suq al-Anad. An army officer quarrelled with a qat dealer over the price, refusing to pay the tax. As a result shaykh Husayn al-Surabi of Bani Mu'adh set up Suq al-Anad in his own territory and beyond the control of the taxman.

The naming of the qat market is significant as Suq al-Anad carries a double meaning. Firstly, to any Yemeni that lived through the 1994 Civil War the name stands for the bloody military battle to capture the large military camp al-Anad, designed by Soviet technicians and equipped with Soviet intelligence hardware. Those who lived through the 70–day war and followed the reports and rumours, including this writer, got the sense that qat and religion played an important role in sustaining the effort to defeat the military might of al-Anad camp. Secondly, al-Anat is Arabic for stubbornness and resistance and the naming of Bani Mu'adh's new qat market is locally interpreted as symbolic for the ability of Sa'dah's tribal population in general, and that of shaykh Husayn al-Surabi in particular, to outwit and resist government control. Since the establishment of Suq al-Anad in 1994 the size and volume of Sa'dah's qat market has shrunk. In a time of economic austerity even buyers from Sa'dah town make the short journey north to Suq al-Anad to benefit from tax-free shopping.

Social Values

> Long live qat, which makes us kind
> and makes us stay peacefully at home
> with our friends. Yemeni Women's song (quoted in Kennedy 1987:79).

Home use   Quist (1990:63) indicates that 60 per cent of the men and 30 per cent of all women – of course separately – participate in the qat session every afternoon. 3–4 hours are thus spent and Yemenis often argue that the qat 'party' keeps the family together and helps to solve conflict peacefully. But qat consumption is certainly very heavy on the family budget. In the 1980s the amount needed per person and day of just medium quality cost between US $10–$30. For the Sa'dah areas in 1997 a person would expect to spend a minimum of US$ 1–2 (YR 150–250) per qat party. During the late 1980s half of the average monthly income was said to be spent on the habit (Quist 1990:64). The same is still true today for the Sa'dah area where the average person without qat trees would spend between YR 3,000 and 5,000 per month, which amounts to half the monthly income of a civil servant. Given the relative high expenses of the chewing habit most farmers will grow enough qat to supply their own daily needs and for special social occasions, such as weddings, when a host is expected to provide large amounts of qat for the family's guests.

Qat and community   The qat session forms the setting in which the bride price is negotiated, as well as a major part of the wedding. During times of sickness or bereavement people come together to chew, often all through the night. Qat is also said to assist religious meditation. After the safe return of pilgrims from Mecca a special qat session is held. Instead of ordinary water used by the chewers to extract the chemical agent (Cathinone) of the qat plant, Zamzam water from the sacred well of Mecca will be passed around, adding a special significance to the qat ritual. But also during the fast of Ramadhan the enjoyment of the qat session, now held after sunset, is considerably enhanced by the intense thirst endured during the day. Qat is chewed to increase the ability to work in the afternoon. Many Yemenis can not afford the luxury of joining the daily qat session in a comfortable sitting room. They will, however, not sacrifice the pleasure of qat chewing and claim that qat diminishes fatigue after the large lunch meal and enables them to do hard work in the afternoon. A stone cutter, for instance, is reported to have said 'I can cut three stones in the morning but I finish 8–10 stones in the afternoon' while chewing qat (Kennedy 1987:79). Many truck drivers who transport goods between Sa'dah and the capital chew qat while driving to keep alert. And this author has met qat packers who every night climb over mountains by foot on most dangerous tracks to reach a feeder road to the qat market. They claimed that only after having chewed qat are they able to set out through the night. Students chew qat while studying for their exams (Quist 1990:64) and Kennedy (1987:12) suggests that even during exams qat is allowed. The often heard argument that millions of work hours are wasted in Yemen every day as a result of the afternoon qat parties is not necessarily true.

Water Use Efficiency

Qat is drought resistant   Qat is a hardy plant, which goes dormant during periods of prolonged water stress (Kopp 1977:32). Once it receives water the tree will almost immediately, after two weeks only, yield a new harvest. In the late 1980s Al Ammar's rocky tracts suffered from a lack of rain. At the same time groundwater levels in the area started to fall to unprecedented low levels, over 150 metres in places and water became a premium commodity. Under these circumstances qat became (and still is) the dominant crop. Many other crops were abandoned as the high returns for qat, combined with the tree's ability to endure high levels of water stress made qat production an economically rational choice.

Irrigation and timing   Tutwiler and Carapico (1981:52) note that, compared with other perishable and high value crops such as alfalfa, water melon and vegetables, qat has the lowest water requirement. This is confirmed by Ward (1998:33) who states that qat 'if correctly watered is one of the most water efficient crops, which coupled with its very high returns make it a very lucrative crop for the farmer.

Over irrigation   Overall, farmers take care not to over-irrigate qat. This is due to the widely held perception that over-irrigation harms qat and affects its quality negatively.

Adjustments to Water Scarcity

As an increasing number of farmers are unable to chase the declining water table and yields have dropped to half of what they were in the early 1980s many now turn to qat as a way of survival. Whereas in the 1980s qat was often seen as an indicator of wealth in the late 1990s for many it has become indicative of poverty. Returns from qat are needed to finance maintenance and repair costs for pumps. Moreover, profits from qat subsidise other farm activities which, although having a much lower economic value, are nevertheless treasured for their social and cultural value.

Changing Attitudes to Qat

Attitudes to the production and consumption of qat are slowly changing. Sa'dah has always been a centre of religious learning. In recent years the region has witnessed a spiritual renewal among the wider population. The influence of Wahhabi thought and practice, propagated by Saudi-trained scholars[7] of Islam is being felt (Haykel 1995:21). In many conversations with people from many different walks of life qat was referred to as the 'tree of the devil' (*shajarat iblis*). The economic benefits of growing qat are perceived as offset by the social obligations this creates, which has led many to conclude that there is no 'blessing' on qat (*ma fi barakah ala al-qat*). Farmers who own large qat fields also tend to entertain daily qat parties, which, by the nature of events, 'eat up' potential returns. Shaykhs and traders are particularly vulnerable. One prominent shaykh shifted from qat to fruit production for this reason. Others are concerned about their children. While they themselves chew qat regularly they do not want to grow it on their farms so that their children may not have 'free access' to the substance and get used to it at an early stage. In contrast to the 'tree of the devil' (*shajarat iblis*) citrus is perceived as a 'tree which is blessed' (*shajarah mubarakah*). Fruit growing is seen as a contribution to health and an activity upon which the blessing of God rests, an honourable gain. The next section will analyse the various values and benefits of fruit production.

Citrus

Citrus trees were not known in the Sa'dah area until the socio-economic changes in the 1970s. One farmer, now in his 70s, mentioned that his father brought two citrus trees back from Taif in Saudi Arabia, which he had visited during the Muslim *hajj* pilgrimage. The trees survived the journey and were planted in their garden but turned out to be extremely sour and only useful for marmalade.

Even by the end of the 1970s perennial crops, with the exception of grapes, pomegranates and qat, had received little attention. For this period Kopp (1981:121) observed that farmers lacked basic knowledge and experience. For example, many farmers thought that apple and orange trees would bear one or two pieces of fruit only. Moreover, advised about methods to increase production, farmers were reluctant to prune their trees, let alone uproot unproductive ones, activities viewed as 'outrageous' and 'barbaric' at the time (Kopp (1981:122).

In the Sa'dah basin much of what has been learned about the production of citrus over the past two decades has come through a trial and error approach. Farmers are very fond of their fruit trees and find it difficult to recognise purely economic consideration, now that profits have plummeted over the past years. Many poorer farmers continue to apply their increasingly limited groundwater to irrigate citrus trees knowing that net returns are likely to be minimal. 'What shall we do, cut down our trees?' The one time trees were cut down was in the autumn 1996 when sentiments raged in protest over a 66 percent increase in the price of diesel, which is essential for running the pumps. Angry Sa'dah farmers cut down truck loads of fruit trees and dumped them in front of the governor's main gate blocking the entry.

Figure 5.10 shows the various values of citrus as perceived by actors in the Sa'dah's basin. The top of this figure points to the economic, political and social factors that drive the production of citrus in the Sa'dah basin. The lower section of the figure indicates some of the main factors, which influence water use efficiency.

Economic Values of Citrus

Prompted by government policy and enticed by economic and political incentives most farmers in the Sa'dah basin shifted to fruit production in the mid-1980s (Chapter 4). This meant investments of three to five years until any financial returns from the first fruits could be expected. Unlike the farmers in the Amran basin (Handley and al-Saqqaf 1996:7) who grow mainly seasonal vegetables and can

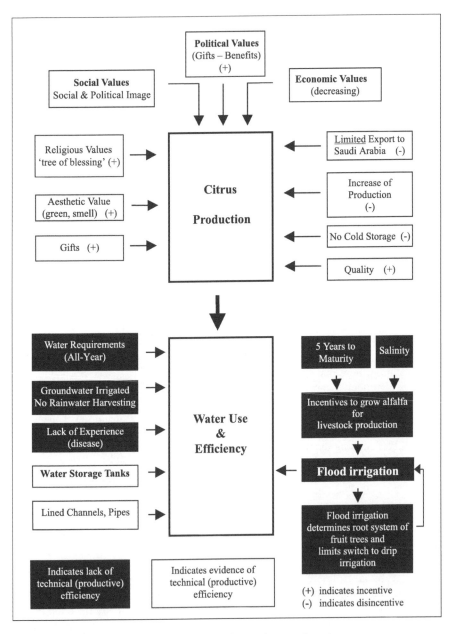

Figure 5.10   Citrus production: values, incentives and water use

therefore respond quickly to price incentives and market demand, the Sa'dah fruit farmers feel trapped by trees they have nourished and cared for over the past 15 years. They point out that the government should accept some responsibility for the dire state many are now in. Repeatedly, the government has promised to open up export markets for their high quality fruits. Instead, demand has not matched supply and net returns have plummeted over the past years.

Shaykhs and traders, especially, responded quickly to demands for cash crops after the import ban for fruit and vegetables was enforced in 1984. Smaller farmers soon followed their example. Sa'dah's farmers pride themselves on the production of grapes and pomegranate (DHV 1993*a*:5) and many shifted to citrus especially, which promised high returns. Price fluctuation, however, has become one of their biggest nightmares. It has taken farmers many years of investment and all those who have mainly fruit trees can not reallocate water resources from one season to the next.

In comparison to citrus, which has seen diminishing returns over the past years, prices for grapes and pomegranates appear to be stable – perhaps explained by the fact that Saudi Arabia will gladly import these two varieties. In some of the traditional agricultural areas grapes and pomegranates are still grown by utilising rainwater harvesting methods. So called *anthari* grapes are 30 percent more expensive than groundwater-irrigated varieties. Moreover, grapes and pomegranates require less water than citrus. Both crops are not irrigated for four months after the autumn harvest while citrus is irrigated all year long. In an attempt to save water a number of farmers have started to shift from citrus to pomegranates over the past few years.

Oranges from the Sa'dah basin are renowned for their sweetness and high quality. At the fruit markets of San'a and Ta'iz they are usually preferred over oranges from Ma'rib, a second major area for citrus production. In spite of their competitive edge producers from the Sa'dah basin face a number of problems.

Markets   Although Sa'dah farmers love their citrus trees, oranges do not figure much in their diet. Markets depend almost entirely upon supply and demand at Suq al-Madhbah, a wholesale market for fruit and vegetables in the capital. Given that there are no cold storage facilities for fruit in Sa'dah, and indeed very limited capacity in the country as a whole, local markets experience oversaturation during the annual harvest season in late autumn when the bulk of perishable citrus ends up at San'a's Suq al-Madhbah (al-Sakkaf et al 1998:14). **Figure 5.11** shows that according to the country's Statistical Yearbook 1995 the area planted with orange trees doubled between 1990 and 1995, from 5,651 to 11,714 hectares. During the same time the production of oranges more than tripled, from 13,896 to 48,731 tons, reflecting both the annual increase of trees and yield of the trees. By 1990 orange trees planted in the mid-1980s were only beginning to bear fruit. By 1995 the trees had reached maturity. Between the harvest of 1994 and 1995 alone orange production almost doubled from 25,991 to 48,731 tons (Republic of Yemen 1996:63).

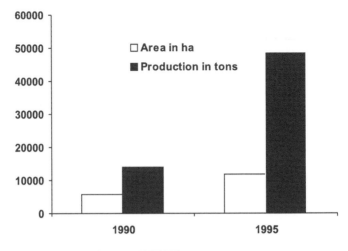

Source: Republic of Yemen 1995:62f

Figure 5.11   (A) Increase of area and production of oranges in Yemen between 1990–1995

Figure 5.11   (B) Fruit from the Sa'dah basin packed for transport to the capital San'a

Reduced returns   Between 1996 and 1997 the price for a *sattel* (5kg) of oranges at the local Sa'dah market dropped from YR 700 to YR 350. This was due to oversupply according to local informants who closely monitored developments at the San'a *suq*.

In the late 1980s when increasing amounts of local oranges first appeared in the urban markets, economic returns provided strong incentives for citrus production (DHV 1993a:5). One farmer, who claims that he was among the first to plant oranges in the mid-1980s, recalled that in 1988 his first fruits were sold at YR 4,000 for a box of 35 kg. During the early 1990s this unit price dropped to YR 3,000. By 1997 one of his boxes would fetch only between YR 1,500 and 2,000 at the wholesale market in the capital where he, like everyone else, takes the fruit. His loss was massively compounded by the rapid devaluation of the Yemeni Riyal (YR) over the same period, from YR 10 to 125 per US dollar.[8]

This farmer has recently tried to export some of his oranges to Saudi Arabia. In late 1998 he took a load of 2.5 tonnes of his prime oranges via the border crossing of Harad, in the Tihamah, to Saudi Arabia. Permission to enter by the Yemeni vehicle was given for one month but the length-of-stay for the driver was restricted to one week. The farmer paid 14% (SR 280 = YR 10,000) in taxes on the estimated value of his load (SR 2,600) and sold his oranges at SR 65 per *sallah* (18kg), the equivalent of YR 125 per kg (US$ 1.0). Back home, at the Suq al-Madhbah in the capital San'a a *sallah* (18 kg) of his quality oranges were traded at about YR 1,000 (YR 50 per kg).

A 1998 economic comparison for exporting oranges to Saudi Arabia confirms that profits could be doubled (AREA, 1998: CA1C). However, a World Bank study has concluded that 'exports are unlikely to grow fast' and that 'regional markets for Yemen's fresh fruit and vegetables are saturated and politically vulnerable' (World Bank, 1998:4). With respect to citrus this opinion is shared by those from the Sa'dah basin with detailed knowledge of trade opportunities with Saudi Arabia. 'Najran is full of citrus', they say. Also, the few who have tried to export oranges to the neighbouring kingdom argue that the Saudis discourage imports from Yemen and put all kinds of bureaucratic obstacles in the way. Only Sa'dah's sweetest and best oranges are permitted to enter. The picture is very different, however, with respect to grapes, raisins and pomegranates from the Sa'dah basin, quantities of which are exported every year to Saudi Arabia.

Roads   Until 2001 no suitable direct road for transporting perishable fruit to the Saudi border existed, in spite of the area's close proximity. The above farmer drove two hours south to Huth, where a newly asphalted road links to the Red Sea coastal Tihamah in three more hours. From there he turned north again to reach the Yemeni border crossing of Harad. Altogether, the journey to deliver 2.5 tonnes of oranges in a small pick-up took several days, a time investment, which is rarely taken into consideration. Such a visit to Saudi Arabia, however, might offer a number of 'fringe benefits' to experienced multipurpose farmers, like the one described.

So far, very few have tried to export their oranges. Most producers from the Sa'dah basin take their products to the fruit market in San'a. During the citrus harvest 60 pick-ups and small trucks daily make the 245 km journey to San'a's Suq al-Midhbah. At the peak of the season the number of trucks can even reach one hundred. This amounts to about a minimum of 145 tonnes of citrus daily (**Figure 5.12**).

The rapid increase in the cost of petrol has increased transportation costs. During the fieldwork period, between 1996 and 1999 the price for petrol tripled from 12 Riyal per litre in 1996 to 35 Riyal per litre in 1999. For one informant, who makes the journey several times a week, the cost for just petrol and oil has gone up from YR 1,000 to 5,000 per round-trip.

Figure 5.13 indicates additional constraints for the marketing of perishable agricultural products from the Sa'dah basin. Along the 245 km to the capital San'a the road crosses the territorial boundaries of Yemen's main tribal confederations four times. Tribal conflict, violation of each other's 'peace' and incidents of blood feud may, at any time, result in tribal check points or roadblocks. Sa'dah farmers taking fruit and vegetables to San'a require daily updates about conditions along the road. They travel armed (as is usual in most northern tribal areas) and in small convoys for the protection of their products (to the market) and their money (on the way back home). In January 1999, for example, the road was blocked to Sa'dah's qat traders.[9] And following the kidnapping, on January 17 1999 of several foreign NGO workers by a Bakil section on Hashid territory, tribal roadblocks were set up by Hashid to control the movements of Bakil who had violated Hashid's territorial space by the kidnapping.

These prevailing socio-political conditions and concerns not only affect the marketing of agricultural products but inhibit economic development of the Sa'dah basin as a whole. They also explain the virtual absence of foreign NGOs working in the area. During the fieldwork period most officials working for multilateral and bi-lateral donor organisations were not permitted by their organisations to travel the road up to Sa'dah.[10]

Political Values of Citrus

The political value of citrus became evident from a frank and open conversation with one leading tribal personality from the Sa'dah basin. From his orange harvest, with a value of YR 2.4 million (US$ 19,200) at the time, he send about one third (YR 750,000, US$ 6,000) as 'gifts' to prominent officials in the government, and to high-ranking officers in the army. The shaykh was trained at the country's leading military academy and holds a considerable military rank. Given his political 'connections' with the army and his desire to provide more arms for his tribesmen he shipped YR 400,000 worth of oranges to a leading actor in the Ministry of Defence. The official reciprocated his generosity by arranging a 'free' licence for the shaykh to import weapons (*silah,* pl. *aslihah*). Weapons, in turn, are important 'gifts' in Yemen's tribal society (Dresch 1993:256, 374). They are often exchanged

Figure 5.12 (A) & (B) Citrus fruit is packaged for transport
It takes a six-hour journey to the capital. With limited availability for cold storage saturation of local markets during the peak-harvesting season explains low returns.

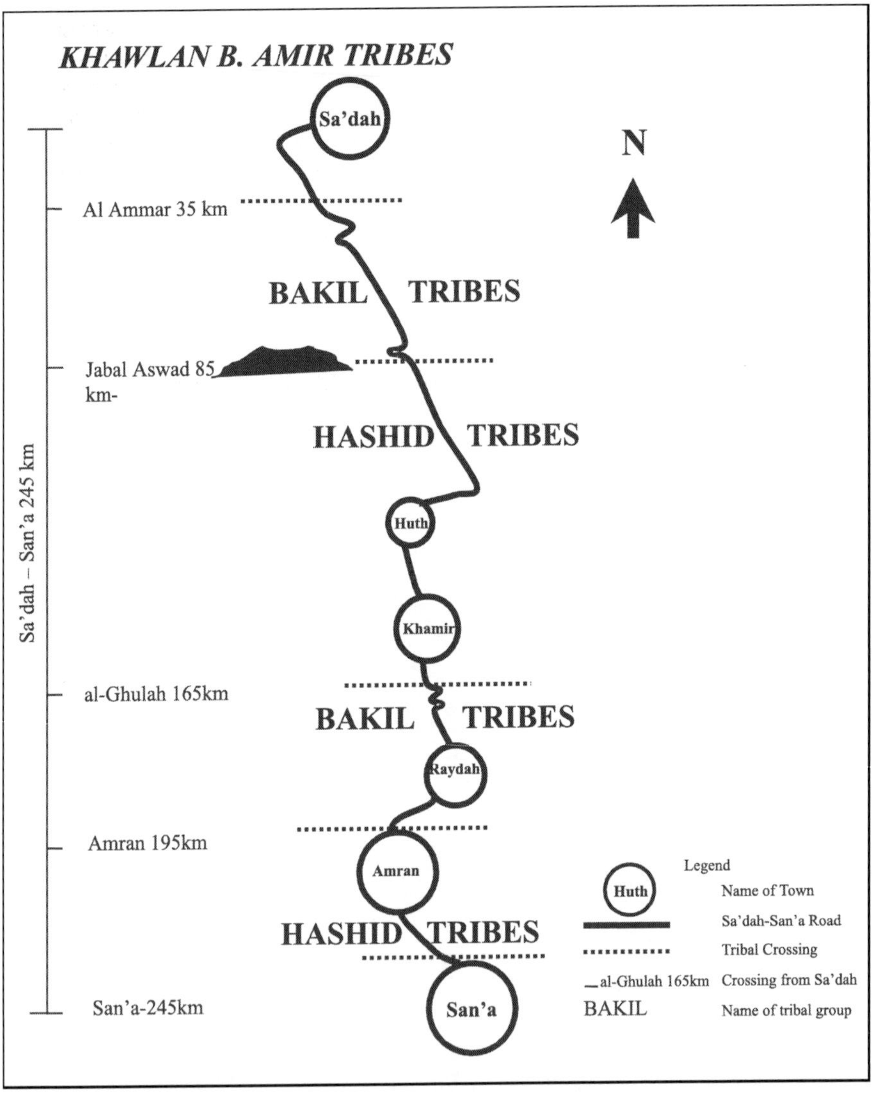

Figure 5.13   Roads and risks: tribes and their territorial crossing points along the Sa'dah – San'a road

to settle blood feuds. They also increase the tribe's bargaining power with the state. Moreover, some shaykhs use small weapons as gifts to 'buy' support and legitimacy. The Sa'dah area is infamous for its weapons trade and has a well-known market where many, especially smaller weapons are openly sold and bought.

The shaykh's main contact in the Ministry of Defence has since been transferred to the Ministry of Transport. The tribe in question controls important territory along a major road and the shaykh was anticipating some 'political' returns from his next citrus harvest.

Many informants emphasised the political value of citrus for shaykhs and traders. The answer of one farmer, who has close contacts with a number of tribal leaders, reflects and represents the perceptions of many others. When asked why influential shaykhs and traders maintained large fruit orchards, he replied that 'most of the fruit is given as gifts to government officials' (*mas'ulin fi-l-hukumah*). One of Sa'dah's leading business entrepreneurs provides another striking example. He has set aside his best citrus orchard for 'gifts'. In conversation he readily recalled the precise number of boxes given, as well as their recipients.

As the cash value of citrus has dropped over the past years pomegranates have also been used for gift purposes. One farmer who is known for his exceptionally good pomegranates reported that certain local officials buy from him in order to give 'gifts' to people in positions higher than them.

Social Values of Citrus

In addition to their economic and political values the production of fruit has a high social values in the Sa'dah basin. According to a socio-economic survey 90 percent of all farmers interviewed in the Sa'dah area had fruit trees (DHV 1993*a*:x). Grapes and pomegranates were the most important fruits before 1984 (DHV 1993*b*:26). The production of fruit has a higher social value. In the late 1970s, when large quantities of apples and oranges were smuggled across from Saudi Arabia, whole boxes were frequently given as gifts to members of a foreign medical team at the Sa'dah Republican Hospital. Now, similar quantities of home-grown oranges and other fruits still arrive at harvest time. On one occasion during fieldwork a trader delivered two boxes with over 30 kg of oranges to the house where the author was staying.

No visit to a farm is considered complete without the obligatory stroll through the orchards to see the fruit trees. During harvest periods the host will select the best and sweetest of oranges, cut them and present them to his guests (**Figure 5.14**). Given that a number of fieldwork periods coincided with the citrus harvest in autumn bags of fruit were also given at the end of each farm visit to show generosity and hospitality.

The 'gift' value of fruit, however, can also discourage smaller farmers from planting an orchard.[11] In one wadi area where no citrus is grown a farmer opted for qat although the fertile soil would have suited fruit trees. Asked about his choice he said:

No one else has oranges. If I were the only one with fruit trees there would be a lot of social pressure to give 'gifts' of fruit to friends and relatives and to people from our community. Not much would be left to take to market. In our area most people have qat so I too planted qat. That way there is no pressure to hand out 'gifts'. It ensures the profit.

Figure 5.14 Visiting the orchard
During harvest season the host handpicks some of the best and sweetest fruits and cuts them open for his visitors to eat. Note also the wooden sticks to support the heavy branches and to protect them from breaking.

Large citrus orchards are perceived as prestigious and for many symbolic of the social, economic and political power of shaykhs and traders. Ownership of a citrus orchard implies the adequate provision of groundwater for irrigation from one or more privately owned wells.[12] Furthermore, for many the production of citrus assumes the economic independence of actors, a desirable goal within Sa'dah's tribal society. 'Sharing a well does not work for citrus' *(hamdiyat ma yaslah li-l-shuraka)* is a perception that was repeatedly voiced.

The prestige value of citrus was also confirmed for some other areas (Dresch, personal communication). Shaykh Mujahid Abu Shawarib of Kharif, whose influence reaches far beyond his own Hashid tribal constituency,[13] started citrus production. People emulate their shaykhs, especially when they are regarded as highly as Mujahid. Soon after many in the area followed their shaykh and likewise planted citrus trees.

Informants reported similar developments for the Sa'dah basin. Tribal shaykhs and traders were the first to get into the large-scale production of citrus in the mid-1980s, at a time when the economic prospects were promising. Many emulated them although it became quite evident that the agricultural 'success' of a shaykh, as one informant put it, 'was not the result of his father's hard work but because of the government'[14] *(mish min ta'b abu wa lakin min al-hukumah)*.

Water Use and Efficiency

Flood irrigation As graphically indicated in Figure 5.10 (bottom right) flood irrigation has been applied for a number of reasons. First, in the mid-1980s groundwater appeared to be available in abundance and pumps were operated at low cost. Secondly, flood irrigation was applied in the citrus orchards to grow alfalfa between the maturing trees. With the loss of much natural grazing land due to the expansion of irrigated agriculture alfalfa had to be produced on the farms. Thirdly, farmers argued that flood irrigation and the alfalfa would decrease salinity in the soil. Consequently, the root system of the trees has developed in response to irrigation method, frequency and supply. Some traders have experimented with drip-irrigation, perhaps in the hope of marketing the idea and trade in modern irrigation equipment. But all those who tried abandoned its use claiming that the mature trees soon suffer from drought stress (**Figure 5.15**). This is a classic example of what Turton (1999) refers to as a solution that has been developed by a technocratic elite without appropriate knowledge of prevailing conditions.

Water requirements Compared to pomegranates and grapes, which do not require water for 3–4 months after their harvest in early autumn, citrus trees need to be irrigated regularly over the whole year. This has increased the demand for water.

Water storage tanks Over the past five years owners of larger orchards exceeding two ha have invested in on-farm water storage tanks (**Figure 5.16**). Two kinds have been observed: a) concrete reservoirs constructed on elevated ground near the pump and up to three metres high with capacities of up to 100 m³ and b) smaller metal tanks with capacities of up to 10 m³ sitting on structures up to five metres high. The forces of gravity facilitate the speedy delivery of irrigation water reducing losses in the unlined channels. Moreover, the irrigation schedule can be timed for the cooler

Figure 5.15 (A) Evidence of drip irrigation (top photo)
(B) Salinity (bottom photo)
As evident from the white patches salinity poses a constraint to drip irrigation; flood irrigation leeches the soil of salt.

Figure 5.16   Water storage tank

evenings when losses from evaporation are reduced. Over 100 water storage tanks are reported for the Bani Mu'adh area alone.

Lined channels and pipes[15]   Farmers in Sa'dah are generally characterised by their openness to new and innovative ideas. Other areas where farmers have introduced measures of technical (productive) efficiency include lining their irrigation channels, using pipes to convey water to the beginning of each row of fruit trees (**Figures 5.17 and 5.18**), spacing out irrigation frequency, and the digging of ditches along fruit trees to supply water more directly to the root system. Drip irrigation is applied consistently in greenhouses of which about 50 exist in the basin.

Comparative Overview of Crops

Tables 5.1 and 5.2 provide a comparative overview of the incentives and constraints of major fruit crops in the Sa'dah basin. It also summarises the relevant findings. It indicates that rainwater harvesting is still applied in traditional communities for the production of grapes and pomegranates.[16] Many farmers continue to take professional pride in the production of grapes and pomegranates in spite of them being a very labour intensive crop (Gingrich and Heiss 1986:98–101)

Figure 5.17   Evidence of technical (productive) efficiency
At shaykh Surabi's farm some irrigation channels have been lined
with rocks.

which perhaps explains why grape production has seen little expansion. According
to one survey grapes are grown on 7 percent of the cropped land in the Sa'dah basin
area (DHV 1993*a*:5). No pesticides are used for grapes and so called *anthari* grapes,
which ripen without supplementary irrigation are valued for their sweetness and
superior quality. Their market value is up to 1/3 above the price for grapes irrigated
by groundwater.[17]

Figure 5.18   (A) & (B) Lined channels and plastic pipes
These measures are aimed to minimise losses.

Table 5.1 Incentives and constraints of major crops in the Sa'dah basin

Crops	Who/Where	Incentive	Constraint	Water	Trends and Prospects
Grapes	Wadis, old agricultural areas	Quality and reputation Cash and subsistence crop Exports to Saudi Arabia No chemicals used Social value (gifts)	Labour intensive Needs expertise	Rainwater and well irrigation Productive efficiency Drought resistance	Demand in Saudi Arabia Continuously high economic and social value High price for rain-fed grapes
Pomegranate	Wadis, old areas	Quality and reputation Exports to Saudi Arabia Storage capacity	Needs expertise Damage from over irrigation	Save 4 months irrigation	Price stability Social value Change from orange to pomegranate
Citrus	Traders shaykhs Walad Mas'ud	Social value (visits, gifts) Religious value (barakah) Economic value (quality) Political value (benefits) Aesthetic value (smell, green)	Disease Lack of experience Saturation of markets Five years to production Water	All-year (every 7-8 days) Storage tanks pipes	Limited export to Saudi Arabia Price fall 1996-1997 by 50%
Apple	New areas Fruit farmers	Economic value Social value	Disease Lack of experience	Saves 4 months irrigation	Declining
Qat	Most farmers Bani Mu'adh Poor farmers	Demand and price Local markets Soil conditions Flexibility of irrigation Dry-season demand Demand in Saudi Arabia 2 harvests per year	Changing religious value ('tree of the devil') fear of disease guarding and theft social obligations loss of benefit	Drought resistant Only when demanded	Religious climate (wahhabism) Poor need to crop qat Expansion to poorer soils Way of adjusting to water scarcity

Table 5.2   Incentives and constraints of major crops in the Sa'dah basin

Crops	Who & Where	Incentive	Constraint	Water	Trends & Prospects
Alfalfa	Near Sa'dah town Most farms Areas with high salinity	Livestock (social value) Livestock (economic value, exports) Improvement of saline soils Grown among fruit trees for the first five years until they bear fruit Markets in Sa'dah town	Water scarcity Lack of livestock Daily labour (marketing)	Water intensive	Increasing with loss of grazing land Increase with rising numbers of livestock in Sa'dah town Increase in eastern saline areas
Sorghum	Small & medium farms 90 % of farmers	Social value Breakfast (with *laban* milk) Fodder for livestock Food self-sufficiency		Irrigated and rain-fed (summer rains)	Strong social and political value as need for autonomy and food self-sufficiency is perceived (crises, war, corruption etc)
Wheat		Eaten with meat (lunch)	Cheap & sufficient imports Need to allow land to lay fallow	Winter wheat needs to be irrigated	Grown only if sufficient rains in late autumn Mostly imported wheat purchased from local markets
Barley		Social value (*hilbah*) lunch	Cheap barley bought from Saudi Arabia for fodder	Mostly winter irrigated	High social value *baladi* food Local barley perceived as superior to Saudi barley which is used only for fodder

Grapevines are also drought resistant and can endure prolonged periods of water stress. Most vineyards in the Sa'dah area receive water allotments from periodic runoff or, in wadi areas, from occasional spates. One grape farmer reported that in 1997 he had applied supplementary irrigation from his well only twice during that season. Indeed, as also in the case of pomegranates, farmers in traditional fruit producing area are aware that too much water during the late maturing period in July and August will harm both fruits. In the case of pomegranates over-irrigation causes the fruit to split open minimising their market value. It is this lack of knowledge regarding the precise irrigation water requirement of both grapes and pomegranates, which in the 1980s lead some new and inexperienced farmers to shift from pomegranates to citrus. Compared with citrus, which needs to be irrigated regularly all-year long the production of grapes and pomegranates offers other substantial water savings. After the grape harvest in August no irrigation water is required until late January when work starts again in the vineyards. Likewise, pomegranate trees need no water for the four months following the harvest in September.

Figure 5.19 shows the crop water requirements (CWR) for selected crops. It indicates that the net CWR for grapes (1093 mm) is substantially lower than for fruit (1297 mm). As both grapes and most pomegranates continue to benefit from supplementary irrigation of runoff groundwater accounts for only part of these figures. Water savings are also apparent with respect to qat, with a net crop water requirement of 889 mm as compared to the 1297 mm of fruit. The experience of qat farmers that excessive amounts of water harm the shrub helps to prevent over-irrigation and therefore saves groundwater. This again is in contrast to citrus where excessive irrigation is applied in order to leach the soil of salt. Moreover, in the Sa'dah basin more qat is irrigated in the dry winter season when prices are high due to the limited amounts of qat available from rainfed areas. During this cooler period losses from evaporation are minimised.

Lastly, vines reportedly produce for up to one hundred years and pomegranates for 40–50 years while orange and apple trees in the area are said to peak at 20 and 10 years respectively. **Figure 5.20** is a graphic perspective and summary of the various values of grapes.

Exports

While the prospects to export citrus to Saudi Arabia appear limited (World Bank 1998:4) grapes, raisins and pomegranates from the Sa'dah basin are highly valued across the border. Even during and after the Gulf Crisis, which put political strain on the relationship between the two countries, the Saudis reportedly allowed raisins from the Sa'dah area to enter their markets. The past few years have also seen renewed export of grapes and pomegranates, according to local informants.

Grapes and raisins produced by rainwater harvesting, without supplementary irrigation, are sweet and fetch high prices. Depending on runoff volume, yields from a one-hectare vineyard can vary. In the mid-1990 farmers in Bani Uwayr reported

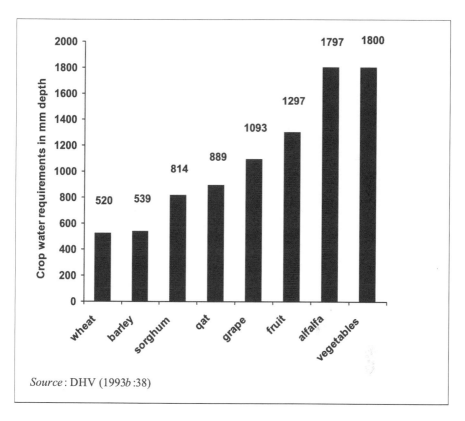

Source: DHV (1993*b*:38)

Figure 5.19 Sa'dah basin crop water requirements

sales of between YR 130,000 in a bad year (*da'if*) and YR 350,000 in a good one (US$ 1,040–2,800) per hectare. During the visit to Bani Uwayr in 1996 farmers mentioned that a trader had just visited the area a day earlier to pick up a load of raisins for export to Saudi Arabia.

A second reason why the economic value of locally produced grapes and pomegranates has remained high is explained by the fact that these fruits have seen little expansion. Unlike citrus, the increase in the production of grapes between 1990 and 1995 has been insignificant, from 142,379 to 150,563 tonnes (Republic of Yemen 1996:63).

Pomegranates, too, have experienced a surge in value. In the Sa'dah area they are increasingly perceived as a precious local fruit (*baladi*),[18] which is associated with traditional customs and values. In the 1980s there was no market for pomegranates

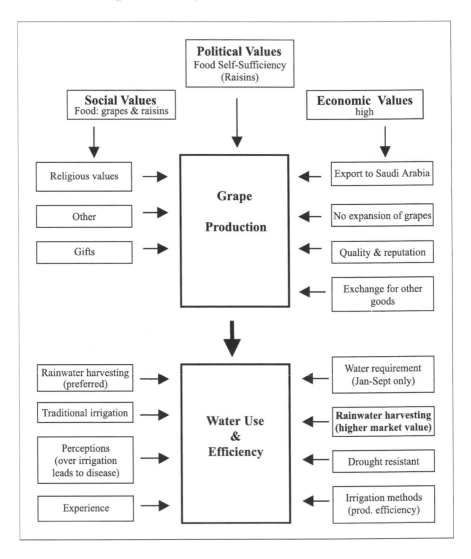

Figure 5.20   Grape production: values, incentives and water use

from Sa'dah in the capital. In the late 1990s local farmers estimated that 80 percent of their pomegranates were taken to urban markets in San'a.

Pomegranates have a lower yield per hectare than oranges but price trends in recent years have compensated for that. At the end of the harvest in October 1996 one kg of pomegranates sold for up to YR 450 at the local Sa'dah market compared to a high of YR 200 for oranges. Average prices for pomegranates during the peak of the season have seen an increase from YR 35/kg in the early 1990s to YR 90/kg in 1997. Over the same period the average market price for oranges has remained at between YR 50 to YR 70/kg.

By the late 1990s some of Sa'dah's traders and shaykhs had started to respond to these changing market signals. Unlike smaller farmers they have the financial capital to shift from citrus to pomegranates. The shift in allocation has, in part, been driven by disease problems experienced with the citrus trees, which meant huge costs for pesticides (Yemen Times Oct 6 1997). A leading local shaykh planted parts of his newly acquired farm with pomegranates. According to local informants he bought the 25,000 *hablah* (62 hectares) of land for YR 50 million (US\$ 400,000) but sold half of it with profit to another shaykh from outside the Sa'dah area. Incidentally, the farm is not in the shaykh's own territory but in the water stressed area of al-Dumayd.

Local grapes and pomegranates have retained their market value even after the recent lifting of the 1984 fruit import ban. The quality of both fruits is thought to be superior to potential imports. Even before the fruit import ban no grapes or pomegranates were smuggled or imported into the area.

Qat and Alfalfa

Qat and alfalfa have been treated earlier in this chapter. In the Sa'dah basin the values of both crops remain linked to unofficial cross-border trade. The main incentives and constraints for production are summarised again in the comparative overview (Tables 5.1 and 5.2.) It is shown that both crops provide coping strategies for farmers in areas that a) experience water stress and b) suffer from the negative effects of salinity. Qat occupies a prominent place on sandy and marginal soils and alfalfa is widely grown between young maturing fruit trees where it is believed to improve the soil quality. The photo in **Figure 5.21,** taken in 1996, reveals that this trend is continuing where young trees are planted.

Cereals

The flatbreads of Yemen, like the country's incredible architecture, were a mirror of each local environment (Alford and Duguid 1995:25).

Of paramount importance to Yemenis, bread is always freshly baked and served warm with meals (Kennedy 1998:25).

Figure 5.21   Alfalfa is usually grown between maturing trees
Alfalfa improves the soil and reduces soil salinity. However, as a
result of flood irrigation to grow alfalfa the trees' root system
develops shallow and far. Farmers report that a later switch to water
saving drip irrigation causes mature trees to suffer from extreme

With respect to cereals **Figure 5.22** indicates that 85 percent of Sa'dah farmers grow
sorghum (DHV 1992:57). The crop's value for the rearing of livestock is one main
consideration for this. Winter wheat, by contrast, is grown less and less as long as
cheap imported varieties are available on the local markets. However, where farmers
still hold rights to runoff and when rains in late autumn occur, as was the case in
1997, many will plant local wheat and barley varieties.

The social values of local (*baladi*) cereals, indicated in **Table 5.3**, are shown in
more detail in the context of food consumption. Bread is the main stable food of the
northern region and the many local varieties testify to its importance. In the average
rural household women members bake at least two-three different kinds of bread
twice every day.

Table 5.3 Cereals: their value and use in the Sa'dah region

	Cereal	Food			Remarks	Other Uses
		Breakfast	**Lunch**	**Other**		
Local	*Sor-ghum*	With Milk (laban) – khubs	With Milk (*laban*) – shafut – lahuh	Long-term storage (*madfan*)	Rural	Livestock
	Wheat	With Fat & Honey (*samn & asal*) – khubs	With meat & broth (*lahm & shurbah*) – ku'ab – khubs	Long-term storage Travel bread (*luqmah*)	Rural	
	Barley		With fenugreek (*hilbah & saltah*) – saltah – muluj		Rural	
	Bakur (local variety)			Long-term storage	Rural Rain-water harvesting	
	'US' Wheat new variety				Rural	Cross-border trade Demand in Saudi (biscuits)
Import	Wheat	Bread Bakeries Markets, Restaurants Urban area	Bread in urban area		Urban Markets	Cross-border trade Livestock
	Wheat-flour		Sweet cake dish With honey *bint as-sahan*		Rural Urban Special occasions (Fridays)	
	Barley					Livestock
	Rice		With meat & chicken		Urban & Afflent people	

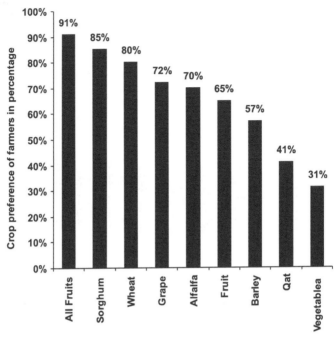

Source: DHV (1993*a*:6)

Figure 5.22   Sa'dah basin: farmers' crop preference
> The survey includes data from farms in some wadi areas at the periphery of the Sa'dah basin. It is noted that less than 10 % of 57 farmers interviewed in the 1992 survey reported no fruit trees. Typically fruit trees occupy $^1/_4$ of the farm. Cereals, especially sorghum for fodder purposes, continue to be very important in the region and are grown by 85 and 80 percent respectively.

Breakfast is usually eaten before 8 a.m. and consists of bread freshly baked from home-grown sorghum and a bowl of sour milk from the farm cow. Household members sit around a bowl of sour milk (*laban*) which is nicely spiced and dip the sorghum bread into the milk bowl. From time to time the bowl is passed around and the milk is shared as a drink. Sorghum bread is eaten with a sour milk dish. It is preferred for breakfast for the perceived strength it provides to do the farm work. Alternatively, or sometimes in addition, bread made from wheat (*burr*), preferably a local variety, is eaten for breakfast. Before it is served wheat bread is broken into small pieces and placed into a preheated stone dish. Liquid butterfat (*samn*) is then poured over the bread and allowed to be absorbed. On special occasions or in honour of a visitor honey (*asal*) will be added to the fat for taste and to sweeten the dish.

Different kinds of bread and cereal-based dishes are the main food for lunch which will usually be eaten between 12:30 and 1:30. Lunch is socially very important. Farmers will come home from their fieldwork, tribal and town markets come to a stand-still and government employees call it a day and make their way home to have lunch with family and friends.

Ku'ab is an important wheat dish special in the Sa'dah area and no lunch will be considered worthy of its name without it. *Ku'ab* is the name given to a wheat dough which is half-baked. Chunks of the dough, the size of a large potato are served in a basket along with the broth from the meat. Smaller morsels, the size of a large walnut, are then taken from the larger piece of dough and the trick is to shape the half-baked dough into a small 'cup' by moulding it skilfully around one's right-hand's thumb. The small cup is then filled with the hot broth. After drinking the broth (*maraq*) the cup is eaten and a new one has to be shaped to drink more broth. The process of moulding and shaping many small cups from dough takes time but also allows for the boiling hot *maraq* to cool.

A dish called *saltah* comes next and takes the middle part of a typical lunch in the Sa'dah area. It could perhaps be described as a vegetable stew but essentially consists of boiling hot broth, which is served this time with fenugreek (*hilbah*), mixed in. According to the season and availability, potatoes or other vegetables will be boiled in the broth before the *hilbah* is added. Whereas the *ku'ab* will always consist of wheat dough, the *saltah* dish will be eaten with barley bread baked freshly just before lunch. Small pieces of barley bread are torn from large round -shaped pieces and used to dip into the *saltah* and scoop out it's contents.

Lastly, the meat will be brought into the room. The head of the household will divide the pieces as he sees fit and fair and distribute them to guests and members of his family alike. The meat pieces are eaten with bread from wheat which, like the *ku'ab* and the sorghum bread, has been freshly prepared for lunch.

In addition, Table 5.3 points again at the trade of virtual water, made possible by Sa'dah's politicised environment. First, a wheat variety called locally 'US wheat' has recently been grown in response to demand in Saudi Arabia where people believe the grain is favoured for making biscuits. Secondly, over the past two decades quantities of cheap wheat imports or free wheat given as aid have been traded across the border. And thirdly, Table 5.3 indicates that cheap barley of Saudi origin is smuggled into the Sa'dah region where it is used to rear livestock.

Comparative Economic Value Between Alfalfa and Citrus

Surprisingly, perhaps, **Figure 5.23** shows that the production of alfalfa compares favourably with citrus given the low price for oranges at the present time according to a recent crop budget study (Crop Budget, AREA Staff Dhamar).

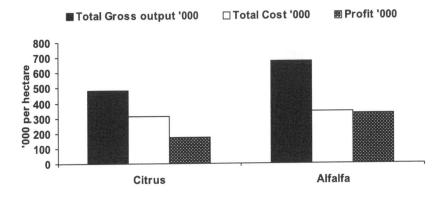

Source: Crop Budgets AREA (1998); World Bank (1993:46f)

Figure 5.23   (A) Crop budgets for citrus and alfalfa. Returns to water in YR per m³ irrigation water

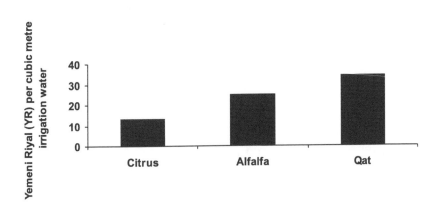

Source: Crop Budets AREA (1998); World Bank (1993:46f)

Figure 5.23   (B) Returns to water in YR per m³ irrigation water

It is based on a low price for oranges, YR 20/kg. Also, for some reason the study assumes two different shadow prices for irrigation water, Yemeni Riyal 5/cubic metre for oranges as opposed to YR 10/cubic metre for alfalfa. Gross irrigation water demand for citrus is put at 2,500 mm compared to 2,400 mm for alfalfa. First, the figure indicates that the total gross output per hectare for alfalfa is 40 percent higher than for citrus. Secondly, it shows that the total costs (input) for alfalfa are lower than citrus (as a percentage of the total gross output), 50 percent (alfalfa) versus 64 percent (citrus). Lastly, the study suggests that under the assumed conditions net profits for alfalfa are above those for oranges, YR 332,000/ha and YR 170,000/ha respectively.

Given the low price for citrus in the study returns to water favour alfalfa with a profit gross margin of YR 25 (alfalfa) versus only YR 13 (oranges) for every cubic metre of irrigation water.

Agricultural Transfer

In conclusion **Figure 5.24** provides a graphic overview of the agricultural transfers between the Sa'dah basin and Saudi Arabia as well as the Yemeni capital San'a. It indicates the time, distance and approach of various routes to the Saudi border. Transfer points for qat are in areas controlled by Khawlan b. Amir to the north-west of the basin, while cereals and livestock are moved mainly through Wa'ilah (Bakil) territory in the east. Wa'ilah also controls the two main roads leading to the official border crossing at al-Buq, Najran. Here raisins, grapes, pomegranates from the basin, and coffee from Khawlan to the west of Sa'dah are exported into the Saudi Kindgom. The road to Baqim, completed since 1981, was finally opened in 2001. The track to Kitaf (Wa'ilah) has been improved since the mid-1990 and provides a fast track for livestock to be moved. From Kitaf rough tracks connect to Saudi Arabia in less than two hours. A longer option is the seven-hour drive through Wadi Amlah (Wa'ilah) used by large trucks. To the west the Harad border crossing at the Red Sea is only 150 kms away but takes up to 8 hours, as the track has to wind its way through the mountainous Razih area. As mentioned above farmers transporting perishable citrus to the Harad border crossing have preferred a long detour through Harf Sufyan, which takes 8 hours.

Fruit and vegetables as well as sand and stones are among the main products and resources transferred between Sa'dah and San'a. The risk due to tribal factors along the 245 km stretch of road have already been referred to earlier in this chapter (Figure 5.13).

Conclusion

This chapter has provided evidence that crop choice and water use pattern are strongly influenced by socio-political and socio-economic values. The data and the

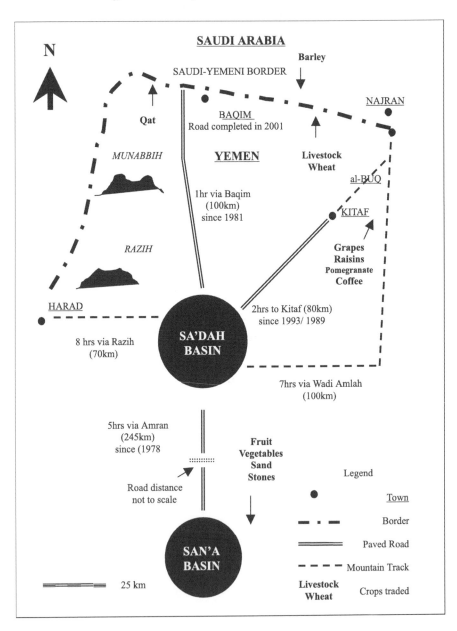

Figure 5.24   Agricultural transfer between the Sa'dah basin and Saudi Arabia

case studies have shown that Sa'dah's politicised environment fosters a political economy, in which skilful actors can thrive by evading government control on one hand and by capitalising on subsidies available through both the Yemeni and the Saudi governments on the other.

Notes

[1] Even in the early 1970s water from hand-dug wells in Sa'dah town was 'definitely very salty indeed', according to medical development workers who came to Sa'dah during those years.

[2] Recent opinions speak of 75 or even 80 percent of farmland in the Bani Mu'adh areas allocated to qat.

[3] This figure includes 'new' areas outside the old walled city of Sa'dah, such as Madinat al-Salam and others (Republic of Yemen ,1996, Population and Housing Census Dec-1994).

[4] Even during the early 1980s qat was smuggled into the Kingdom, often from areas much further away from the Saudi border. Morris (1986:123) reports that '[s]ome foolhardy drivers are said to have successfully smuggled qat into Saudi Arabia inside inner tubes or watermelons or hidden under the chassis of their vehicles. However, this is a highly dangerous form of smuggling, given that the sale and consumption of qaat in the kingdom is punished with even greater severity than that of alcohol.'

[5] On January 30, 1991, the small Saudi border town of Khafji, some twelve miles from the border with Kuwait was overrun and occupied by Iraqi forces. The town's was retaken by an Arab force composed of Saudis and Qataris a few days later.

[6] This information was obtained from a visitor to the Asir region.

[7] The village of al-Dammaj, just seven km south-east of Sa'dah town has become the centre of a large group of Islamic students from around the world around the Saudi-trained *hadith* scholar Muqbil al-Wada'i (died in 2001). His centre in al-Dammaj was visited several times during field work. The Yemen Times (10 August 1998) considers shaykh Muqbil 'to be the leading theoretician of the [salafiyya] movement in Yemen.' For a recent interview with the *salafi* leader see Yemen Times 30 July 2000. See also Weir, S. (1997) A clash of Fundamentalisms: Wahhabism in Yemen. Middle East Report (July – September 1997).

[8] A similar drop in profits was experienced by citrus farmers in Sicily/Italy. Between the 1960 and 1980s citrus famers on the island made 'big' money (ca 4,000 pounds sterling on one hectare [6 *tumulo*], according to local informants). However, since the mid-1990 the citrus market has been 'dead' due to imports from North Africa. Some citrus farmers have been able to survive by shifting to olives. (personal communication with a native of Sicily.

[9] Usually the qat trade is not affected by tribal incidents or roadblocks, which increases the crop's comparative advantage. For example, during the civil war in the 1960s 'Al-Ahnum's qat trade with the plateau went on throughout most of the war, under agreements guaranteed in common by men who on other grounds were at daggers drawn...(Dresch 1993:245).

[10] Between December 1998 and February 1999 there were eight separate incidents in six governorates involving 35 foreign nationals from America, Austria, Holland, Germany, Italy and the UK (Yemen Times 6 February 1999). Very few of the western professionals resident in the capital San'a who were visited and interviewed for the purpose of this research between 1996–2000 had been permitted by their agencies to travel to the Sa'dah area.

[11] Serjeant 1995 VII:32f) mentions the custom to give first fruits as gifts to travellers and that it is considered shameful to ask money for such. Those of the vine are called *subuh* of grapes. The heads of *dhurah* are given to the poor as a first fruit offering.

[12] Grohmann (1933:30) reports earlier for the area of the Hadramaut 'where one does not say that such and such person owns land but such and such owns a well.'

[13] In one recent example shaykh Mujahid negotiated a truce between the eastern tribes of Murad and Al-Tohaif over 400 square kilometres of land near Ma'rib. The conflict had left 18 people dead and 26 wounded (Yemen Times, 40, Oct. 98).

[14] Travelling the fringes of the eastern desert near the oasis of Ma'rib in 1995 nomads pointed out the large groundwater-irrigated citrus farms in the area, many of which are owned by well-known personalities linked to the state and military apparatus.

[15] The spread of schistosomiasis (bilharzia), which is endemic in many areas has driven many farmers to install pipes for conveying irrigation water. Open water sources and standing pools provide a suitable environment for snails, the disease's intermediate host. Over the past two decades increased awareness on these issues has also come through Egyptian teachers familiar with bilharzia.

[16] Fresh grapes were a favourite food of the prophet Muhammad and raisins are considered an aphrodisiac and believed to have medical value (Varisco, 1994).

[17] According to a 1997 study (Ward et al. 1998:18) grapes were the second most profitable crop in the San'a (after qat) as far as returns to water is concerned.

[18] A caricature in the Yemeni newspaper al-Thawra 12 May 1985 supports *baladi* fruits. It shows a *baladi* banana and a *baladi* orange kicking out imported (*mustawrid*) fruit cans (Weiter 1987:420).

6 Adjusting to Water Scarcity

There is no moving creature on earth
But its sustenance dependeth on Allah

Sa'dah farmer's response to water scarcity
quoting the Qur'an (Surah 11:6)

Introduction

This chapter will explore the comparative social, economic and political resourcefulness of selected actors and communities to adjust to water scarcity. In addition it will indicate the socio-political and socio-economic potential and constraints at work in a politicised environment such as the Sa'dah basin, to implement water demand management strategies. While there is evidence of economic (allocative) and technical (productive) water efficiency it will nevertheless become clear that increasing water scarcity has resulted in resource capture and has led to a degree of ecological marginalisation. At the same time, there are examples of adaptive capacity to resist resource capture by powerful actors. While this is encouraging Chapter 6 will indicate that tribal-political change in the Sa'dah basin over the past two decades appears to inhibit community participation and the formation of water user groups to address the challenges posed by groundwater mining. Finally, it will be argued that adaptive capacity is obscured by legal pluralism, a contradiction within Yemen's water legislation, which seeks to reconcile established belief systems as enshrined in customary and Islamic law with a modern water legislation based on principles of sustainability and equity.

Figure 6.1 suggests a conceptual framework for Chapter 6. At the top it indicates that impact and perception of groundwater depletion are affected by different factors; social, economic, political, ecological, religious. In turn, responses and solutions to water scarcity, whether actually implemented or perceived, depend on impact and perception. Responses and solution can include supply management or water demand management, i.e. productive and allocative efficiency. Alternatively, economic development and sectoral reallocation of water provide remedies to water scarce environments. However, each of these options is governed by the prevalent socio-economic and socio-political realities at the time. The potential and constraint of these variables determines the level of groundwater depletion as indicated by the two arrows that return to top box, groundwater depletion, thereby closing the circle.

Factors Influencing Groundwater Allocation and Management:

A Conceptual Framework

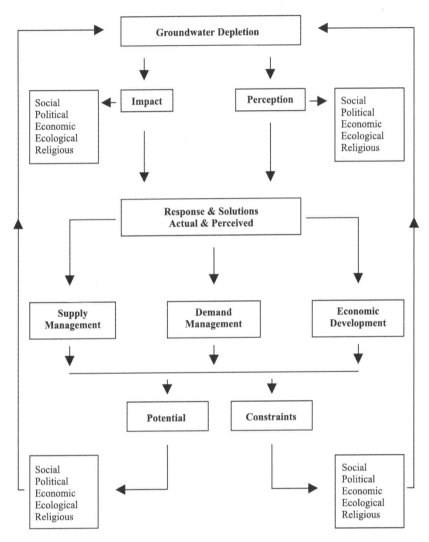

Figure 6.1   Factors influencing groundwater allocation and management: a conceptual framework

Perceptions

The fact that groundwater is usually referred to as *bahr* (sea) indicates the dominant local perception that groundwater is 'limitless' like the water of the sea.[1] Since the early 1990s this perception has slowly been corrected by an increasing awareness (*wa'y*) of groundwater deficits (**Figure 6.2**). However, the same word *bahr* was still used in a letter, sent to this writer in 1998, by a farmer desperately seeking financial help to 'reach the *bahr*' in his area where the tide (to continue with the metaphor) had gone out fast and did not seem to come back in.

Town water supply in Sa'dah town is inadequate to meet demand and a number of small tankers supply customers from bore holes owned by farmers outside the town. Satisfying the need for drinking water (*haqq al-shurb*) is a religious merit. Until 1993 a religious family, which had earlier migrated to the Sa'dah basin from the east, allowed private water vendors to fill up their 3.5 cubic metre tankers from their farm well free of charge. Around 1993 their well could no longer cope with the numerous water trucks that filled up daily to supply customers in Sa'dah town with water for domestic needs. Their own situation demanded that they had to stop the practice. They irrigate several hectares of fruit trees and, at the time, their extended family amounted to 95 adults and children sharing one large house. From 1995 on the family has had to deepen their two wells two times, from 100 to 200 metres and again to 250 metres.

Supply Options

One entrepreneur from the Bani Mu'adh area has developed an air pump, which, the inventor claims, is able to lift groundwater from depths of 450 metres at a much lower cost.[2] The pump has attracted the interest of the Yemeni media and was shown on TV. The family owns heavy equipment for the drilling and repair of wells. With fewer farmers in a position to afford expensive repairs developing a cheaper alternative for lifting groundwater from increasing depths is the solution envisaged by these local entrepreneurs. In 1999 the family said that ten air pumps had been built and installed, another 25 were ordered, mainly for other areas in Yemen.

Figure 6.3 shows the supply strategy of a poor farmer in the central area of Bani Mu'adh. The old hand-dug well near his house has been dry since the mid-1980s. Diminishing volumes in his own well have forced him to neglect his vineyard. After only 30 minutes of operation the volumes from his pump ceased and he had to wait some hours for the well to recharge. The farmer decided to see whether he could help the process of recharge. **Figure 6.3 (A)** shows the excavation by a bulldozer near his well. He hopes rainwater runoff to fill the depression thereby applying recharge directly to his well.

The time for this came in 1997. During much of the 1980s and until the mid-1990s rainfall in the Sa'dah basin had been sporadic and scarce. Unexpected heavy

Figure 6.2   Dried up
(A) Hand-dug well (top photo) (B) abandoned vineyard (bottom photo)

Figure 6.3   (A) Hoping for well recharge

Figure 6.3   (B) Well repair
Since the early 1990s most farmers have had to add three pipes (six metres) every year to keep up with dropping groundwater table.

rainfall occurred in the Yemeni highlands during the dry winter season (*shita*) of November 1997 when average rainfall is usually at its lowest measuring 3.3 mm for San'a and even less for Sa'dah (Van der Gun 1995:3.2–2, 3.2–7). From the 250 mm of rain that fell in the capital San'a between January and October 1997 almost one third or about 100 mm of rain fell during the week 19–26 October alone (Al Lindholm, November 1997, private communication).

Dark rain clouds covered the Sa'dah basin during much of the same period. People from the village of Sawdan (Bani Mu'adh) reported three flash floods (*sayl*) during a single day. Rainfall throughout the region also brought the trade in qat to a stop as people had more immediate concerns on their mind, to repair damage caused to their mud houses and to save their grapes which during this period are being dried into raisins on the flat roofs.

Heavy rainfall also occurred during Yemen's rainy season in 1998, so much so that the World Bank's adviser for the water sector wrote that 'no one believes about water scarcity in Yemen' (Chris Ward, World Bank, personal email). As also observed in southern Africa periods of drought can provide a window of opportunity as it shifts public perception from supply options to demand management. Good seasons of rainfall, in turn, can result in the public perception that the crisis is over (Turton 1999:39 n.38).

Constraints to Drip Irrigation

Farmers in the Sa'dah basin take great pride in the hard work of farming. Asked about drip irrigation systems one farmer emphatically stated that he perceived such measures as a lazy option. He said that 'flood irrigation is hard work and more honourable than drip irrigation... all they [farmers with drip irrigation] have to do is pump the water up into the tank and from there it will run down'.

In one case reported a young innovative farmer had started to experiment with drip irrigation. His father, however, did not think highly of such innovation. When the son was away for a few days his father reverted to flood irrigation.

Many of those who started to grow cash crops in the early 1980s were inexperienced and have progressed by the method of 'trial and error'. 'Most things around here are done randomly (*ashwa'i*), there is little planning and forethought, everyone just does as he pleases' or phrases to that effect are often heard. Also, groundwater appeared limitless two decades ago and people felt little need to take productive efficiency into consideration.

One farmer near the qat market of Suq al-Khafji planted one row of qat between every three rows of citrus when he first started. He flood-irrigates both crops together, every 8–10 days, according to the requirements of his citrus tree and 'the qat does not seem to mind'. Initially, he also had pomegranates but was too inexperienced with this fruit and replaced them with citrus.

One shaykh who planted 15 more hectares with citrus in 1997, a time when it was well known that groundwater was being mined, has not introduced drip irrigation. Asked why water saving technology had not been considered for the newly planned orchard his farmers responded as follows:

> there is no need for drip irrigation here, we have sufficient irrigation water from our four deep wells. Fortunately for us, the water table leans (*mayyal*) our way and we are blessed with good quantities of groundwater.

The answer is indicative of the way many people in the basin think and supports the notion of the 'tragedy of the commons', benefits to the individuals accrue at the cost to the community. Moreover, the notion of scarcity is understood not in absolute terms but in terms of the inability of actors to access a resource. Some informants, even after they had already run out of water, maintained the view that it was only a matter of ability, not sustainability. Ability, in turn, is perceived as predetermined, a belief which helps people to accept their lot – *al-qawi ya'ish wa al-faqir yamut* (the rich will live and the poor will die) was a phrase heard repeatedly during conversations.

Figure 6.4 shows the newly planted citrus orchard of a 'strong' man (*qawi*). Irrigation water is conducted through unlined channels. To benefit from some of the losses, however, tomatoes have been planted in the irrigation channels.

The number of wells, too, is perceived as indicative of actors' financial ability and economic power. This became evident during many conversations. Saying that such and such person 'has four wells on his farm' conveys an image of power and not one of scarcity.

Images of power are also created through the technology used for drilling a new well. Over the past years a number of powerful rotary drilling rigs have operated in the Sa'dah basin. Drilling by this method costs about YR 3,000 per metre (US$24) as opposed to only YR 1,000 per metre (US$8) when employing hammer drill equipment.

Figure 6.5 shows a rotary drilling rig drilling a new for one of Sa'dah's leading shaykhs during 1996. The sheer noise during operation is overwhelming and makes it impossible to have a conversation with those involved. In this case the well was drilled to a depths of 400 metres and the whole process lasted just over 24 hours. Rotary drilling equipment was also used in another case that was witnessed during 1999, at the home of bin Ja'far, shaykh of Sahar and the main rival to the shaykh mentioned above.

In addition to the drilling equipment employed, the kind of pump used also reflects ability and conveys an image of power. By 1999 a few actors were known to operate submersible pumps (*ghattas*). One shaykh was able to stop all his six older pumps on one of his farms after he started to operate a submersible pump in his newly drilled well. One or two more submersible pumps are operated in al-Dumayd, the area afflicted by groundwater over-abstraction.

Figure 6.4   Irrigation through unlined channels
Irrigation inefficiency benefits tomatoes, which have been planted in
the unlined channels at a large farm of a Sa'dah shaykh.

A number of the farms in the Sa'dah basin are now owned by absentee landlords,
a fact which inhibits measures of technical (productive) efficiency. In such cases
local people are often employed to live on the farm, guard it and irrigate the fruit
trees. There is little incentive for these workers (*umal*) to conserve irrigation water.
In fact, they derive some benefits from flood irrigation. Flood irrigation allows them
to supplement their income by rearing a few livestock on the weeds and residues and
sometimes on the alfalfa they grow in between the trees (**Figure 6.6**). One 8 hectare
farm supplied by four wells near the Sa'dah airport belongs to a trader from al-

Figure 6.5   Drilling a new well with a powerful *dawaran* (rotary) drill
At the time the photo was taken (October 1996) a depth of 110 metres
had been reached with no sign of groundwater. To ensure adequate
volumes the well was drilled to about 400 metres. It is powered by a
submersible pump.

Ahnum (Bakil) who reportedly hardly ever visits. Workers irrigate the citrus trees
which occupy 2/3 of the land with the remaining 1/3 being allocated to cereals.
Neighbours say 'he leaves everything to his workers. In some cases involving
absentee landlords, however, resident farmers do not hesitate to express that they are
not pleased when water is allocated to trees with no productive value. A close
relative of the former Deputy Prime Minister, the man owns a farm in the Sa'dah
basin. People from Al Uqab (Bani Mu'adh) owned the land but sold it to a fellow
from Al Ammar who then sold it on to al-Anisi. The owner's choice of trees
included many 'exotic' and non fruit-bearing varieties, which rather upset many of
the local farmers who feel that 'water should not be wasted on trees without food
value (*fa'idah*)'. For the same reason the production of fruit with nutritious value is
preferred by many farmers in the Sa'dah basin over the production of too much qat,
which is perceived as lacking 'benefit' (*fa'idah*) even by those who chew most days.

Figure 6.6   Sheep grazing in a citrus orchard
Due to loss of grazing land most livestock are now kept on the farm.
Flood irrigation of trees creates a desired side effect; sheep can graze
on the residues, weeds and grasses that grow from excessive irrigation
as seen on the photo.

Constraints to the Adoption of Technically (Productive) Efficient Methods

Pump-sharing arrangements necessitate communication and can foster co-operation
leading to increased community participation, as has been argued elsewhere in
Yemen (Vincent 1990:24). However, pump-sharing can also inhibit water use
efficiency as evident from a case witnessed during the sorghum harvest in the
autumn of 1996. When a pump, shared by several farmers in one of the traditional
farming areas just south of Sa'dah town, broke down during the summer of 1996 it
took the shareholders some time until they had all come up with the money to pay
for the needed repairs. By the time the pump was running again, farmer Salih was
well aware that the yield of his sorghum crop would only amount to half the average

yield due to the missed irrigation cycles. Under different circumstances Salih would have abandoned the crop. But since he had been forced to pay quickly for his share of the repairs the other sharecroppers did not permit him to stop irrigating his sorghum. He felt forced to use his allocated time-share of water for his retarded sorghum crop.

The same 'tragedy of the commons' outcome has also been observed in other areas where poorer farmers share pump and well. 'If one shareholder does not take his water when it is his turn (*dawr*) he will lose it, he cannot take it another time.' This is why most people will make use of their water share in some way even when the resource is not used efficiently.

Productive Efficiency: Known About and Adopted by Some

Many farmers in the Sa'dah area are generally characterised by their openness and innovative ideas. In one rare case it was observed that a farmer had dug ditches along the sides of his fruit trees to supply water more directly to the root system (**Figure 6.7**). The method is known as sub-surface irrigation and has a number of advantages. First, it reduces weed growth by applying the water directly to the roots of the trees, in this case pomegranates. Secondly, losses from evaporation are lower because the surface area directly exposed to sunlight is smaller. Thirdly, surplus water replenishes groundwater faster and lastly, sub-surface irrigation encourages the roots to grow deep instead of spreading wide and shallow, as is the case with flood irrigation.[3] Remarkably, no one had instructed the farmer about sub-surface irrigation, nor had he observed it anywhere else. He simply tested a method that he developed in his own mind. His pomegranates are of fine quality and guarantee him high returns at the local market.

Allocative Efficiency: a Response to Water Shortage

Wadi Sabr / Bani Mu'adh

Falling groundwater levels have prompted farmers to reduce the area under pump irrigation by half and only irrigate higher-value cash crops. In the Bani Mu'adh area, for example, one farmer now irrigates only 500 *hablah* (1.25 hectares) of his 1000 *hablah* (2.5 hectares) plot. His first well dried up after only eight years of operation, in 1987, when a new well was drilled on the opposite side of the farm. Qat and citrus take equal shares of 250 *hablah* (0.65 hectares) on the lower half of the farm. The upper half of the farm is now left to depend on uncertain runoff. If there is adequate runoff during the spring season (March/April) sorghum will be planted there. If no runoff occurs during the spring season but only later during the summer rainy season (July/August) a local variety of barley (*sha'ir bakur*) will be sown, which has a

Figure 6.7   Evidence of technical (productive) efficiency

lower water requirement.[4] If the summer rains come late the farmer will plant *burr* (winter wheat) on half of the farm. This case indicates that allocative efficiency is present. Availability of imported cereals (virtual water) allows the farmer to apply his diminishing volumes of water from his pump to the higher value cash crops, qat and citrus.

Al Ammar: Qat as an Example of Economic (Allocative) Efficiency

The topographical characteristics of Al Ammar have been mentioned in Chapter 3. According to their shaykh, the tribe of Al Ammar numbers between 9,000 and 10,000, of which 2,000 are counted as arm-bearing men. Al Ammar had a large weekly tribal market until the late 1970s when the government demanded a share of ten percent in taxes from the increasing turnover of goods traded there as a result of the emerging cross-border trade.[5] Al Ammar's market is still famed for its qat but has, over the past two decades, lost its former market position to Suq al-Talh, the largest market in the Sa'dah area. The six old wells around the village of Al Ammar stopped yielding any significant amounts of water by the late 1980s. Groundwater levels have dropped to between 150 and 180 metres, confirmed by the experience of farmers who need between 50 and 60 pipes (3 metres per pipe) to reach groundwater levels. However, for flow volumes to be adequate wells have to be drilled to depths of around 400 metres.[6] The drilling of modern tube wells in Al Ammar started in the early 1980s, when YR 45,000 (US$10,000) would pay for well and pump. Now, no less than YR three million ($US24,000) will buy the same. Drilling is still cheap, costing YR 2,500 for a metre, but the price for equipment is beyond the reach of farmers now. In addition to the cost for pipes (YR 18,000 for a three metre piece) farmers in Al Ammar need the more powerful and costly 35 hp pumps, compared with the 28 hp pumps still used by farmers in the Sa'dah basin. The high capital and running cost explain why some pumps in the area of Al Ammar are shared by up to 20 farmers. It is, therefore, not surprising that qat is the major crop in Al Ammar. As seen in Chapter 5 qat has a high cash crop value and tolerates plant stress, qualities, which make the plant highly allocative efficient.

Qat from Al Ammar (qat *ammari*) has a reputation for high quality and is famed throughout Yemen. Qat *ammari* fetches highest prices in the country's qat markets. People travelling from Sa'dah to the capital will usually make a short stop at Suq Al Ammar to purchase freshly picked *ammari* qat for the journey. They may even buy some extra bunches of the shrub, which they will try to sell with profit at their destination.

For most farmers in Al Ammar qat is the main source of livelihood. The high returns from qat help farmers to cope with water scarcity, population growth and with a lack of alternative economic activities. Over the past decade and especially since the 1990/91 Gulf Crisis qat production is said to have increased by about 40 percent on land adjacent to Ammar's villages and in areas near the qat market. Interviews, and visits to the area of Al Ammar, including Wadi Madhab, where the tribe has large communal land holdings, allow for the following conclusions.

- In areas where wells are deep qat has gradually replaced local cereals.
- In some wadi areas with shallow wells qat has replaced cereals on poor soils. Moreover, qat cultivation has extended onto new areas characterised by poor soils.

- Subsistence agriculture is continued on good soils with relatively easy access to groundwater.
- Large communal landholdings in Wadi Madhab, a tributary of the Jauf graben, allow Ammaris to shift to qat near their villages while continuing cereal production and subsistence agriculture in the distant wadi, where wells are shallow and seasonal spates occur.

Figure 6.8 (A) shows the upper section of the Al Ammar's Wadi Sharamat. On the higher plots grapes are still grown in traditional ways by utilising runoff from seasonal spates compared to **Figure 6.8 (B)** which gives a view of the wadi's lower section where qat has replaced cereals. The marginal soils that produced only poor yields of sorghum now produce high quality qat. Qat grown on these plots is irrigated only every 3–4 months which compares favourably with the 15–20 days irrigation cycle of sorghum. Qat is picked between two and three times per year and in 1997 *Ammari* farmers claimed cross profits of YR 200,000 (US$1,600) from a small plot of 25 *hablah* (0.06 hectares). The reputation of *ammari* qat as well as demand across the border at a particular moment in time may account for the high value.[7]

In spite of the high value of *Ammari* qat subsistence agriculture remains important. **Figure 6.9 (B)** shows the middle section of Wadi Sharamat. In the middle section of the wadi a few shallow wells provide water for irrigation, mainly alfalfa and sorghum. Farmers are aware that returns from qat subsidise many of these lower-return activities *maksab al-qat yughatti takalif al-ashya al-ukhra* (returns from qat cover the costs of other things). However, notions of food self-sufficiency have remained strong among old and young Ammaris alike, especially during the 1990s. As one young man put it, 'the political history of our country provides sufficient evidence to all our generations that we must remain food self-sufficient' (*al-zuruf al-siyasiyyah allati marrat bihi al-bilad ithbadat lil-ajyal darurat ijad al-iktifa al-dhati*). Both, the 1994 civil war and the military stand-off with Eritrea in 1995 sparked fears of an emergency situation (*khawf min azmah*). They also renewed concerns about food self-sufficiency – 'we usually store our own grain (*habb*) in underground cisterns and buy imported grain from the market for our daily needs, but we try not to 'touch' our own supplies.' However, food self-sufficiency is much more of a cultural ideal than an economic reality and some farmers from Al Ammar openly stated that imported cereals, which they buy at nearby markets, meet up to 90 percent of their total demand.

A closer look at Wadi Sharamat reveals some other important characteristics of the area. Firstly, settlements are built at the edge of the wadi to allow for all the fertile soil to be worked. Even the position of their cemeteries reflect the high value of their scarce agricultural land. Their dead are buried on the steep and rocky flanks of the wadi. Secondly, agricultural expansion has reached its limit in Wadi Sharamat. Outside the wadi bed no agriculture is possible, given the lack of soil and water. Land scarcity in the upper Al Ammar region has been a push factor for some

Figure 6.8   Wadi Sharamat
(A) Upper section of Wadi Sharamat (top photo) with mainly
vineyards compared with (B) lower section of Wadi Sharamat (bottom
photo) where qat is the dominant crop.

Figure 6.9   (A) General view of Al Ammar near suq
Note that small plots are around the dwellings, in contrast to Wadi Sharamat (pictured below) where dwellings are at the edge of the wadi so not to occupy valuable soil.

Figure 6.9   (B) General view of Wadi Sharamat

people to buy land in the Sa'dah basin. However, Al Ammar's large territories in Wadi Madhab provide coping strategies for present and future needs.

Wadi Madhab lies about 20 kms to the south of Suq Al Ammar. At lower altitudes malaria is endemic in the wadi, which explains why few Ammaris have chosen to live there permanently.[8] The section of Wadi Madhab claimed by Al Ammar stretches for many kilometres but only in the middle section can a few hamlets be found. The families living here own agricultural land (*mal*) of about 1000 *hablah*/each (2.5 hectares) near their dwellings. There is a problem over runoff. Families cropping a small plot of agricultural land claim rights to runoff from large territories of grazing land, in some cases up to 200 hectares. When privatisation of Madhab's common land was first discussed in the early 1980s Ammaris rejected the solution by the Islamic scholar on the basis of which the tribes in the Sa'dah basin had honoured claims to runoff rights. Applied to Wadi Madhab it meant that the few resident families would come to own large tracts of the wadi leaving the majority of Al Ammar's population with a small remainder. In the end Al Ammar reached their own consensus to solve the land-runoff right issue. It is based on the principle of equity and, importantly, also considers the needs of the next generation. In brief, in 1982 the lower section of the wadi was equally divided among all *Ammari* males. Every male, including male children and babies born on that day acquired private ownership of 50 *hablah* (0.125 hectare) land in the wadi. By 1997 the large remainder of land in Wadi Madhab had not yet been privatised. Ammaris want to wait a little longer so that the children of the next generation will, like them, be able to claim an equal share of land in Wadi Madhab. It must be stressed, however, that non-tribal people living in Al Ammar are excluded from a share in Wadi Madhab. In one case land scarcity has driven a number of *sayyid* families from Al Ammar to the Sa'dah basin.

On some of Al Ammar's communal land in Wadi Madhab rainfed farming of sorghum can be undertaken by any tribal member. Only a written permission is needed from the shaykh specifying the location and size of land to be farmed. The fact that Ammaris are able to grow most of their sorghum for food and fodder on rainfed tracts in Wadi Madhab allows them to shift to high-value cash crops, especially qat, near their villages. The shift to economic (allocative) efficiency in water use has also been possible because it has not meant a reduction of livestock. Households maintain between 15 and 20 sheep and goats. Livestock are fed on *dhurah* fodder (*alaf*) trucked in from Wadi Madhab. In addition, livestock are also taken to the grazing areas in Wadi Madhab, especially after the *dhurah* harvest to feed on the stubble.[9]

In conclusion a number of lessons can be drawn. Farming communities in Yemen do respond to the onset of water shortage. First, measures of economic water use efficiency are a direct response to water scarcity. Secondly, runoff rights have not determined control over natural resources nor led to resource capture. This is because it is in the interest of the majority of Ammaris to redistribute their common land resources in a equitable way. Thirdly, qat has enabled farmers in Al Ammar to cope with diminishing groundwater. Overall, evidence suggests that there have been equitable and possibly sustainable land and water tenure responses. There have also been water allocation and management responses, which reflect awareness and the capacity to act on evidence of water shortages.

Qat Risks

While qat is an ideal cash crop for water-stressed environments, field observations during January 1999 made clear that qat production in the Sa'dah basin is not without its risk. In many of the central areas of Bani Mu'adh, near al-Anad qat market, qat is now grown on over half of the irrigated area. It is also common for farmers to allocate 2/3 of their land to qat production. Typically, this could amount to 1,000 *hablah* out of 1,500 *hablah* land to qat production (2.5 out of 3.75 hectares).

Qat does not tolerate frost.[10] At an elevation of 1,900 metres the Sa'dah basin can experience night frost during the winter period. One night of frost during January 1999 destroyed a large percentage of the qat harvest in the Sa'dah basin and qat farmers, from the Bani Mu'adh area in particular, lost millions of Yemeni Riyal in revenue. Localised pockets of frost do occur every 3–4 years but the last cold spell remembered as affecting the entire basin was 14 years ago.

The losses to Sa'dah's qat farmers in 1999 were particularly high because profits during the dry winter season are highest when little qat comes from the rain-fed areas. One qat farmer had just agreed to sell his production at a value of YR 1.2 million (US$ 9,600) to a qat trader who was going to have it picked the next morning. Straight from the field the qat was to be taken across the border. When the trader approached the field early the next morning the entire crop had withered. In anger the farmer threw the money back through the window, pulled a blanket over his head and was not seen for many days, or so the story goes.

Figure 6.10 shows the qat farm of one informant after the frost. The frost has 'burned' (*ihtaraq*) the qat and turned the leaves brown. The roots of the trees are still alive but the branches have to be cut back, a loss of at least one year in revenues. This qat farmer had recently irrigated his crop. Although aware of possible frost damage he decided to risk waiting with the harvest. The value of his qat would have doubled from YR 150,000 to YR 300,000 (from US$ 1,200 to 2,400) within a week. In his case the losses were also high. Half of his 4,000 *hablah* (10 hectares) is allocated to qat, the other half to citrus. Some religious people were quick to point out that the frost was a punishment for all those not paying taxes (*zakat*) on their huge revenues from qat.

It needs to be stressed, however, that qat is no longer symbolic of wealth. In many cases qat is indicative of poverty as it allows poorer farmer to survive. As one informant stated, 'if we did not have qat we would be much worse off'.

Figure 6.10   (A) and (B) Qat fields destroyed by a few nights of frost
(January 1999)

Runoff, Resources and Riches in Mahadhir

> There is perhaps no literate Yemeni who is not convinced that the subsurface of his homeland is a vast Aladdin's cave brimming with mineral riches and rare gems (Stookey, 1978:261).

The area of al-Mahadhir, located about ten kilometres south of the Sa'dah basin, provides a poignant illustration to the above quote. The story also indicates the extent to which surface water runoff rights facilitate resource capture.

Within the span of only a few years a formerly poor family have turned into millionaires by quarrying rocks out of land over which they hold runoff rights. A few kilometres to the east of al-Mahadhir's main village, hidden in the barren and rocky outcrops lies an abandoned grape vineyard. Until a few years ago seasonal runoff from the surrounding slopes rainfed the grapes, producing a sweet but meagre harvest. Agricultural development in these rocky hills is constrained by a lack of fertile land and by the depth to groundwater, a fact attested to by a number of tube wells in the area that are left abandoned.

In the early 1990s adjusting to scarcity took a sudden turn for the impoverished family. In desperation they started to quarry the rock on their runoff slopes. At first the rocks were sold to builders and in nearby Sa'dah town. Word about the brilliant white rock spread quickly and soon there was demand in San'a. The houses of the rich and the powerful in the capital were being built with rock from al-Mahadhir.

By 1997, this development had radically transformed the life of this family. Indeed, the economic benefits of this single enterprise have been felt throughout al-Mahadhir.

Eight heavy bulldozers, at the value of YR eight million each, quarry enough rocks to fill six heavy trucks per day. At the cost of YR 100,000 (US$800) per truck load at the quarry its value increases to US$150,000 by the time it reaches the capital. The business creates livelihoods for a number of families from al-Mahadhir; eight drivers at YR 40,000 salary per month; about 30 workers (stone cutters, guards, tyre repairmen etc.). Traditionally, these employees are provided with lunch, which includes lots of meat (several sheep are slaughtered every day), and with qat (with a reported value of about YR 20,000/day) to keep the workers' morale high during the long afternoon hours. As a result, the quarry enterprise has boosted economic development in al-Mahadhir's main village, al-Qabil, which is about seven kilometres away. The number of butcher shops has increased to six. In addition, twelve mechanical workshops have opened, mainly to service and repair the heavy equipment used in the quarry. Now, even people from Sa'dah town come out to al-Mahadhir to have their tractors and other machinery fixed. By late 1997 a new market area had been built.

Ideas that promise financial returns spread quickly. Others in the Sa'dah area have tried to make money from quarrying land. Few areas, however, match the alleged quality of the brilliant white and soft rock quarried by the family described above. However, those in the business of transporting stones from the Sa'dah area to the capital have found a way to get the higher price paid for lower quality rock. With their trucks almost loaded to capacity with lesser quality rocks from elsewhere they run up to the quarry of the al-Mahadhir family to have a few higher quality rocks placed on top of their load in the hope that buyers in the capital will not notice the difference.

Figure 6.11 shows the family's quarrying business during a visit in 1997. As a result, the vineyard no longer receives sufficient runoff and has dried up (**Figure 6.12**). Rights to runoff from a small plot of less than one half of an hectare have given the family the customary right of ownership of the slopes surrounding the small vineyard. Discovering that their rock had value turned these former have nots into rich entrepreneurs in the course of only a few years. What is important about this case is the fact that such 'overnight' success stories shape the perceptions of local people vis-à-vis the potential value of their natural resources. News about such discoveries helps to explain why many others prefer to hold on to their runoff rights in the hope that their land will, some day, reveal hidden treasures. Furthermore, the case helps to explain why influential shaykhs and traders engage in resource capture. And lastly, the unexpected turn of this formerly poor family strengthens the belief that God can, miraculously, supply new sources for livelihoods when other resources, such as groundwater, have become exhausted.

The Water Stressed Area of al-Dumayd: Land Privatisation and Conflict

Figure 6.13 shows the area of al-Dumayd in the tribal territory of Walad Mas'ud (Sahar). A satellite image indicates the extent of irrigated agriculture in al-Dumayd in 1998. Between 1982 and 1992 alone groundwater levels dropped by over 40 metres (DHV 1992:44; Fig.4.20). This case study will briefly look at the consequences of groundwater mining in al-Dumayd and investigate some of the coping strategies available to farmers and communities.

Al Humaydan is a small village in al-Dumayd. In the early 1980s its population did not exceed 200 (males). At the time most adult men, about 50, worked in Saudi Arabia. By 1994 the population of Al Humaydan had increased to a total of 1,358 (685 male and 673 female; Population and Housing Census Dec. 1994:324). After returning from Saudi Arabia in the mid-1980s these 50 men got involved in agriculture. Families with large landholdings often sold parts of their land to finance the agricultural development of new irrigated farms. In turn, land sales in al-Dumayd attracted many tribal people from other areas to invest in irrigated farming.

Al Humaydan's common grazing land (*mahjar*) amounted to a sizeable 50,000 *hablah* (125 hectares) before it was divided up (*taqsim al-mahjar*) in the early 1980s. The people of Al al-Sayfi, a small village to the east of Al Humaydan, claimed the runoff rights (*haqq al-sabb*) from this large 125 hectare area to irrigate their 1,500 *hablah* (3.75 hectares) of agricultural land (*mal*). Conflict broke out when privatisation of Al Humaydan's common land was first considered in the early 1980s, around the time the import ban on fruit and vegetables was enforced. Three

Figure 6.11   (A) The family quarry in al-Mahadhir (top photo)
(B) Up to six trucks per day with a total value of YR 500,000 (US$ 4,000) are being loaded for transport to the capital San'a (bottom photo)

Figure 6.12 **(A) The vineyard of the al-Mahadhir family left to dry up with a few dwellings in the background** (top photo)
(B) Discarded rocks from the quarry against the background of a vineyard (bottom photo)

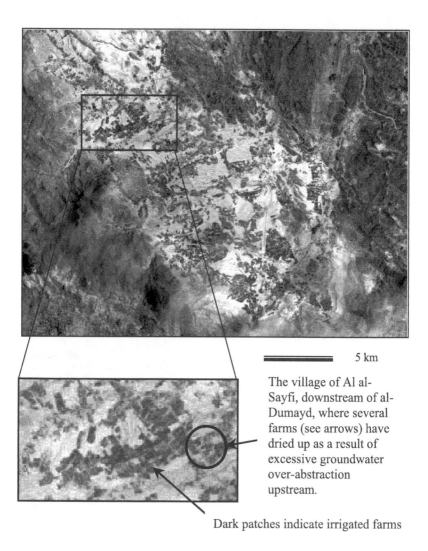

5 km

The village of Al al-Sayfi, downstream of al-Dumayd, where several farms (see arrows) have dried up as a result of excessive groundwater over-abstraction upstream.

Dark patches indicate irrigated farms

Figure 6.13 The water stressed area of al-Dumayd
Since the mid-1980s landowners in al-Dumayd include people from many tribal groups, a fact, which makes co-operation over water management and allocation difficult. As a result of over-abstraction upstream farmers in Al al-Sayfi have run out of groundwater. Date of satellite image: August 1998.

families from Al Humaydan claimed a larger share than the rest and refused to accept the solution proposed earlier by the Islamic scholar on the basis of which half of the common land was equally divided by the number of males. The government was approached to help mediate but the villagers were referred back to customary law (*urf*).

Eventually, daggers (*janbiyya*, pl.. *janabi*) were put on a heap in a symbolic act to indicate that all were determined to solve the land issue peacefully.[11] Mediators were then sent to the families, three sheep were slaughtered in front of their homes, and they were asked to accept the tribal consensus of the village. 'We have decided to give you 3,000 *hablah* (7.5 hectares) in compensation on top of your share in accordance with the Ijri proposal (*ala hasab qarar al-Ijri*)' they announced.

Peace was restored and the common land was privatised. In the process the people of Al Humaydan managed to buy back 12,000 *hablah* (30 hectares) from Al al-Sayfi's share.

The 1982 well inventory indicates the existence of about 30 wells in the al-Dumayd area, most of which were drilled after 1980 (Gamal, N. et al 1985:65f; Appendix 2). Groundwater development in the area was boosted by a) the fruit import ban in the early 1980s and b) the willingness of the Al Humaydan people to sell parts of their land to outsiders. By the late 1990s locals estimated that the total number of wells in the are of al-Dumayd could be several hundred.

Figure 6.14 shows the supply management scenario of a farm in al-Dumayd, which is characteristic for this area. It is a story of trying to chase the rapidly diminishing groundwater resources. Starting with one tube well in 1984 the farm has now exhausted five wells. The table inserted in the above figure indicates the depths of each well, year of drilling and how often it has been deepened. Pump number five is the latest attempt to pipe adequate amounts of groundwater water from a distance of three kilometres away.

Farmer Ahmad comes from a trading family. Dealing in weapons (*asliha*) his father made a small fortune during the Yemeni civil war in the 1960s. During the 1970s the family continued in cross-border trade. In the Al Humaydan area they were the first to equip their old hand-dug farm well with a small diesel pump. In the mid-1970s pump equipment was not available locally. In order to purchase the required parts Ahmad and 20 tribesmen from Al Humaydan had to travel as far Jidda in Saudi Arabia. Pump and pipes were transported south to the Yemeni mountains where the roads stopped. Here, the equipment had to be loaded onto camels for the last stretch of the journey across steep ravines and wadis. On numerous occasions and where the tracks became too steep the long water pipes had to be carried by the men on their backs.

In 1984, when the Yemeni Government banned the import of fruit and vegetables, farmer Ahmad went into joint farm venture with one of Sa'dah's 'big' traders. Their large farm combined an area of 7,000 *hablah* (17.5 ha), the trader owning 4,000 *hablah* (10 hectares) and farmer Ahmad claiming the remaining 3,000 *hablah* (7.5 hectares). With the help of the trader they were able to purchase 3,000

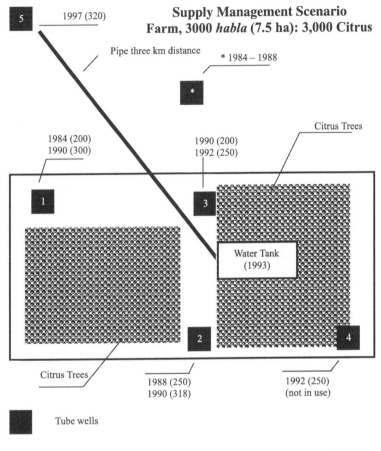

Note: 400 *habla* = 1.0 ha (1 *habla* = 25 m²)

Figure 6.14 Supply management scenario: 7.5 hectare farm planted with 3,000 citrus trees

citrus seedlings from government stores. The seedlings were planted on one section of the farm taking up an area of 3,000 *hablah* (7.5 hectares) or one *hablah* per tree. The large remainder of the farm was left uncultivated. It was agreed that the eastern half of the citrus plot would be owned by the 'big' trader, leaving farmer Ahmad with the 1,500 trees on the western end of the farm. One deep well was drilled to about 200 metres in a northern corner of the farm. Initially, it supplied enough groundwater to irrigate all 3,000 seedlings.

Trouble started four years later in 1988 when the citrus trees started to produce and when it emerged that the 'big' trader's section was more productive than farmer Ahmad's. During the first four years of maturing Ahmad had provided most of the labour and he felt justified to demand that the combined returns from all the trees be split equally between them. When the trader rejected this, conflict resulted between the two families. It was resolved only through the mediation of the tribes' shaykh. Ahmad paid the trader some compensation in return for keeping the 3,000 *hablah* planted with the citrus trees. In return the trader now claimed ownership of the hitherto uncropped 4,000 *hablah* (10 hectare), including the well, which was located on his plot. Pump and motor, however, were claimed by farmer Ahmad who dismantled the equipment, mounting it onto a borehole that had been drilled to 200 metres depth in 1984 but which had not been used.

Soon after, however, it became apparent that the one well could no longer meet the demand for irrigation water. A second well was drilled in 1988, this time to 250 metres while only two years after, in 1990, his first and second well were both deepened from 200 to 320 metres to chase the fast-declining groundwater table and to make up for the gradually diminishing volumes of groundwater in his pumps.

Drilling beyond 250 metres proved a technical challenge at the time and most of the locally available drilling equipment was inappropriate to penetrate through the hard rock. Owners and operators of drilling rigs would often refuse to drill beyond these depths as they frequently experienced breakdowns. The drilling head would bury itself into the hard rock, snap and break.

Farmer Ahmad had two further wells drilled in 1992, each reaching initial depths of 250 metres but one of them has never been used and may now not be deep enough to extract profitable amounts of groundwater. By 1997 the combined volumes of his three operating pumps could no longer meet irrigation demand for his 3,000 citrus trees and farmer Ahmad was looking for new supplies.

Adding to the four existing wells on the citrus farm was deemed unproductive. To find additional supplies of groundwater a fifth well was planned three kilometres away on a piece of land bought from a neighbouring tribe. In 1997 farmer Ahmad contracted a local drilling-rig operator requesting the new well to be drilled to 400 metres. Over the past few years improved drilling rigs have been brought into the Sa'dah basin from Saudi Arabia to meet the new challenges which supply management poses. At a depth of about 370 metres the drilling head snapped and broke and the operator refused to drill deeper.

'In these days one well up there is like two down on the citrus farm' maintains farmer Ahmad. Current yields from well number five are higher than from all his other wells combined. But capital costs for drilling, equipment and water transfer are considerable. In addition, farmer Ahmad had to negotiate the rights to pipe the water to his citrus farm, over three kilometres away. The water pipes had to run subsurface and were allowed only to traverse through uncultivated common land. He was fortunate that his community did not object. Since the new well is closer to the mountains and therefore on slightly higher ground it allows for the water to gravitate down through the pipes to reach the distant farm.

Looking at farmer Ahmad's background and present circumstances it becomes clear that the large citrus orchard is sustained by other incomes. His father was a well-to-do trader who could afford to marry four wives. Ahmad has three brothers and three half brothers. Although Ahmad's father died six years ago the three brothers have not split their share of land between them but continue in joint-venture activities. The older brother has control over the finances and does the planning. The three families own a petrol station at the main road, deal in money exchange (*masrafah*) and engage in the trade of goods (*bay wa shira*). In addition to the citrus farm described above they have another fruit orchard of 1,200 *hablah* (4 hectares) citrus and pomegranates plus two hectares of land on which they crop cereals.

Socially, the brothers have also done well. Recently, farmer Ahmad married one of his daughters to the brother of the tribe's shaykh. Moreover, he enjoys the company of a tribal shaykh from the immediate border area. Ahmad's father had six boys from four wives. Now, the number of male descendants has grown to 75. Livestock numbers have also increased. Ahmad's father used to keep 40 sheep. Now, each of the brothers raises about that number of livestock mainly on groundwater irrigated alfalfa and weeds that are a welcomed by-product of flood irrigation.

Farmer Ahmad is a devout man. He has built a small mosque on the farm where he and his hired day-labourers can pray. Asked about how to address the groundwater problem in light of his children's future needs, his religion gives him hope. He answered, 'We see to this day and do our best until we die, God will take care of the next generation'. Like many others in the Sa'dah basin he too perceives the government as corrupt and bad and therefore will not turn to officials for advice – *la naqrab al hukumah*. At the same time he admits that the area of al-Dumayd poses enormous challenges for community water management, expressed as follows:

> We should not have sold our land to all these people from other tribes. We don't know each other and we don't trust each other, there is hardly any co-operation between us, it will be difficult to achieve a consensus to reduce groundwater abstraction.

His view is also shared by his adult son who adds that 'some of these newcomers to the area are not 'honourable' people, some have been involved in dubious activities and corruption but we found out too late; otherwise we would not have welcomed them here.'

The water-stressed area of al-Dumayd provides an interesting example of the constraints to community co-operation. Since the mid-1980s landowners there include people from the two main tribal confederations Hashid and Bakil with their numerous subsections as well as from the tribes of Khawlan b. Amir with their main subsection, Razih, Munabbih, al-Mahadhir, Kkawlan and Juma'ah. With the exception of a few individuals these new landowners have not changed tribal affiliation by moving among the Sa'dah tribes. They share no history of co-operation with their host communities. Kohler, too, makes this important point by saying that the decentralised structure of the tribal system inhibits the exchange and flow of information necessary to address urgent water issues at the appropriate scale. In the eastern Ma'rib region, for example, down-stream users said that dialogue with farmers up-stream was not possible because they were from another tribe (Kohler 1999:138).

In addition, a primary reason for many farmers from other areas to move to the Sa'dah basin may have been to break free from the need to share and co-operate over scarce and uncertain supplies of surface water in their highland home territory. The legitimate right of a landowner to abstract groundwater on private land is clear to all these farmers. Kohler (1999:138) makes this point very powerfully by reproducing the answer of farmers in the east of Yemen. Asked whether there was a forum for discussing water problems farmers said:

There is nothing to discuss because we have all switched to groundwater irrigation.

Similar responses have been given by farmers in the Sa'dah basin most of whom feel that current interpretations of customary and Islamic law leave little room for discussing restrictions on abstraction rates and well drilling. These realities pose challenges to the formation of local initiatives and user groups in such areas. A fresh Islamic interpretation vis-à-vis groundwater could provide a window of opportunity (Lichtenthäler 1999:19f). However, as long as the currently held belief is not challenged, giving landowners open access to limited groundwater resources, there is little direct conflict between those who can and those who can not exploit the resource any further, as evident from the coping strategies below.

Coping Strategies

al-Dumayd

The land of farmer Ahmad's absentee neighbour, the 'big' trader, has rapidly been developed since 1988. A total of eight pumps now irrigate the 4,000 *hablah* (10 hectares) citrus and apple orchard. The trader family owns between six and eight more farms in the area. The immediate consequences of such resource capture have surfaced over the past few years and lead to tension. Groundwater abstraction from a

500 metre deep well by a submersible pump resulted in a number of wells in the vicinity drying up. When the farmers protested, the owner of the submersible pump consented to operate his pump only by night allowing his neighbours to abstract available volumes by day. But the farmers affected are aware that a 'big' neighbour has no obligation to share these resources.[12]

Ahmad's farm, described above, provides evidence for this. The farm of his immediate neighbour and fellow village member was abandoned because it dried up. Unlike Ahmad his neighbour lacked the means to pursue supply options. In fact, it is highly likely that abstraction from the numerous wells on Ahmad's farm plus that from the eight wells on the farm of the trader caused the well of this third man to dry up. Trying to keep his farm in production the owner tried to sell a piece of disputed land he claimed elsewhere. It only got him into further trouble. Given the declining groundwater levels no one was prepared to share or sell water. Impoverished, he has not sold his main farm yet but might be forced to do so soon. Meanwhile his sons earn a living as day labourers on the citrus farm of their neighbour, farmer Ahmad. Other farmers affirmed the sense and sensibility of such life tragedies by reciting the following proverb:

Al al-Sayfi: Response to Extreme Groundwater Depletion

>   man ja ba wa man shib ishtara
>   One who starves sells, and
>   One who had his fill buys

The following example gives application to the above proverb. One of the communities adversely affected by groundwater mining in the al-Dumayd area is Al al-Sayfi. The village is down-stream from the concentration of large irrigated farms in al-Dumayd, which might explain why many wells have dried up in Al al-Sayfi.

Demand for sand for the building boom in the capital San'a has provided temporary coping strategies for some farmers. By 1999 the reported cases of farmers selling the sand underlying their farms in the al-Dumayd area alone had increased to five. Sand from the Sa'dah basin is valued for its quality, reflected by the 'willingness to pay' a higher price for the commodity in the capital San'a, 245 kilometres away. Farmers in the Sa'dah basin became aware of the value of their local sand by seeing some of their dry wadi beds being dug up.

Wadi Alaf: Mining Sand as a New Livelihood

The area where Wadi Alaf crosses under the main road, and before it enters the Sa'dah basin, is just one example of this practice. However, in some cases, the extent of mining of what would be regarded a common property resource has led to conflict. In Wadi Alaf there was a fatal accident when a child drowned in one of the deep craters that had filled with water after a seasonal flood. In another incident competition over this resource resulted in shooting and the practice of mining sand in the wadi was stopped by the local shaykh.

On private land, however, mining sand is the right of the owner. In areas suffering from extreme water stress, such as al-Dumayd, mining the sandy layers of one's farm has become a strategy for survival. In fact, depending on quality and volumes, some farmers were surprised to learn that profits from sand (*bathah*) were higher than the value of their farm. Financial returns from the first farm in the area to mine sand reportedly came to YR six million (US$ 48,000) whereas the value of the farm's land would not have exceeded YR 1.5 million. For comparison, his neighbour earned YR 100,000 during the same year from the production of citrus and apples on a similar size plot. In another case it was said that sand sales enabled a farmer to build a new house whereas he would not have been able to do so from his agricultural sales.

Figure 6.15 shows the practise of sand mining on a farm in Al al-Sayfi. The owner was forced to stop farming when his well dried up in the mid-1990s. Fertile soil has been bulldozed to the edges of the farm allowing heavy machinery to dig up the sandy layers. By 1997, when the farm was first visited, the crater had already reached a depth of about ten metres and large trucks drove right down to the bottom to be loaded up with sand for the San'a building boom.

Figure 6.15   (A) The remainder of a farm in Al al-Sayfi

Figure 6.15   (B) Loading up sand for the building industry in Sa'na

The owner of the land (*sahib al-ard*) and the owner of the mechanical loader (*sahib al-shaywal*) usually enter some kind of profit splitting arrangement. Mining one's land in this way allows the farmer to keep ownership of the land. Also, farmers argue that agricultural production could be restored, if irrigation water became available. If not, they express the hope that the ground might reveal some other valuable raw materials, such as experienced by the al-Mahadhir family described earlier. And then, there is the option to move the soil to areas where groundwater levels are believed to be better. This is done in the area of Harf Ja'far where, as in al-Dumayd, groundwater levels have fallen by over 40 metres between 1982 and 1992 (DHV 1992:45). Actual depth to groundwater is now over 100 metres.

Harf Ja'far: Soil as Marketable Commodity

Figure 6.16 shows the area of Harf Ja'far on the 1998 satellite image. The photos in **Figure 6.17** were taken in January of 1999. The bottom one (B) shows a truck transporting the soil from one of Harf Ja'far's 'old' agricultural fields (*mal*) to a farm that is being developed in an area where groundwater volumes are believed to be more promising. The photo at the top (A) presents the result, two former plots where the top layer of 1.5 metre of soil has been dug up and taken away.[13]

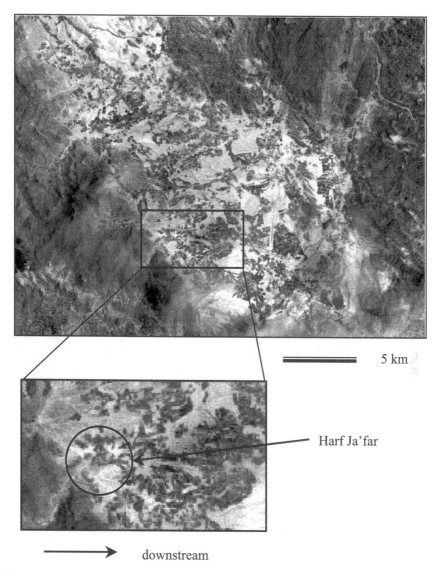

5 km

Harf Ja'far

⟶ downstream

Figure 6.16 The area of Harf Ja'far
where groundwater levels have dropped by over 40 metres between
1982 and 1992 (DHV 1992:45). Coping strategies for farmers in Harf
Ja'far include selling and/or moving fertile farm soil to new locations
where groundwater can more easily be accessed.

Figure 6.17 **Fertile soil from Harf Ja'far is dug up and transported from 'old' fields to 'new' fields where groundwater levels are more promising** (top and bottom photo)

The value of water became apparent during a visit to the village of Harf, the home of shaykh Dirdah bin Ja'far, traditionally the shaykh over all the Sahar tribes. The area suffers from water scarcity and most families, including that of shaykh Dirdah, have moved to live on their newly developed farms. Surprised by the visit and in an attempt to show hospitality to the visitor, a woman ran into her house and soon returned with a gift of five litres of water presented in a former oil container (**Figure 6.18**).

Agriculture is often subsidised by other sources. 'It is a shame to leave the land fallow' was a sentence heard often and people describing themselves as farmers may hold a number of other jobs at the same time. Those working in government offices

Figure 6.18 The social value of water in Harf Ja'far

are usually back home in time for lunch, at 2 p.m. Out of eight farmers that were met working on their land one afternoon seven held other jobs in the morning. One had returned from Libya especially to help with the annual sorghum harvest. Farmers often suspected that there was no net profit *(fa'idah)* from most of their farming activities. In most cases, they did not do a balance sheet comparing expenditures and incomes. When asked why they did not, the men in one gathering answered that they were hesitant to do so since the numbers would probably indicate that they were losers *(khasranin)*. 'Comparing farm expenditures and incomes will only reveal that we should leave some of our crops altogether.' The social pressure to keep the land tilled has been noted by Dresch who states that:

> Land is kept in production wherever possible (it is still reckoned 'untribal' not to do so), although the labour required makes no sense in terms of the cash economy; one could for example, buy wheat cheaply, but the produce of one's own land is reckoned 'better'[14] (Dresch 1993:307).

Moreover, in the aftermath of the Gulf Crisis there are few alternatives outside agriculture although some individuals display a remarkable ability to combine a multiplicity of livelihoods. One individual, for example, from an area close to the border holds a government post in his hometown where he spends half of the week. For another two days he manages to work in the Saudi city of Najran and for the remaining part of the week he travels south to the Yemeni capital San'a where he does some other 'business'.

Exporting Virtual Water to Saudi Arabia

The nature of cross-border activity has changed over the past decade and a half. To travellers visiting Sa'dah during the 1980s, at the peak of smuggling, the place was booming with activity.

> The old Sadah of the Zaydi imams is now, like most Yemeni towns, ringed by a belt of sprawling development. At first sight the place seems to consist of mechanics, oil changers and *bansharis* [puncture repairers]. All the paraphernalia of transport made it look like one huge truck stop (Mackintosh-Smith, 1997:78).

Now, in the aftermath of the 1990/91 Gulf Crisis smuggling is mainly conducted by people walking across the border and taking only small items. The extent to which trade in qat with areas across the border provides coping strategies has already been mentioned in Chapter 5. Elsewhere, it has been suggested that Yemen serves as a transit point for the trafficking of hard drugs to lucrative markets in Saudi Arabia and the Gulf (Yemen Times 30 June 97), a 'livelihood' which in turn poses a considerable risk to life.

In the mid-1990s a new wheat variety was introduced in the Sa'dah basin where it is known as *burr shami* (northern wheat), *dhurah safra* (yellow sorghum) or *burr amriki* (American wheat), because it is believed to be an American variety. Informants reported that 'American wheat' had been grown in the southern areas of Yemen for some time. Sa'dah farmers started to introduce it in response to demand in Saudi Arabia. Yields compare favourably with other local grains, about one *qadah* (20 kg) per *hablah* (8 tonnes per hectare) and returns are high due to demand in Saudi Arabia. In late 1997 one *qadah* (20kg) of 'American wheat' was selling at the al-Talh market between YR1,000 – 1,400 (US$ 3,520 – 4,480 per hectare) compared with only YR 500 per *qadah* for the local *dhurah bayda* (white sorghum). According to many local farmers this new wheat variety is produced solely for Saudi markets. There are hefty benefits for those engaged in smuggling the commodity. Reportedly, the value of one *qadah* (20 kgs) has more than doubled, to Saudi Riyal 100 (YR 3,300) once the sacks reach the Saudi market. No one could say with certainty where this kind of virtual water ends up or why the Saudis prefer this particular variety but many believe that 'American wheat' is used in the Saudi biscuit industry. In Yemen virtual water flows in mysterious ways (Mitchell 1995:3), usually uphill 'towards power and money' as also observed in other Middle Eastern states and in California (Reisner 1993:296; Allan 1999a:1).

Buqalat: The Degradation of Waqf Land

Buqalat is an area just east of the old walled city of Sa'dah (**Figure 6.19**). Much of it is land associated with religious endowment (*waqf*). However, over the past two decades tribal people from the arid plateau of Bani Uwayr have moved in and established some larger farms. Problems associated with declining groundwater levels have been exacerbated by high levels of salinity. In Buqalat coping strategies of farmers have included growing alfalfa for the market in the near-by town of Sa'dah, and turning the soil into clay bricks for sale. Moreover, since the mid-1990s an increasing number of agricultural plots have gone out of production.

Figures 6.20 shows one section of Buqalat known as Bir al-Hadi (the well of Imam al-Hadi) because the entire area used to be irrigated by that well. On the 1.5 hectare (615 *hablah*) of land 62 percent had gone out of agricultural production between 1995 and 1997, when the area was visited. Included in this are two plots of pomegranates (12 percent), which have been left to die. What is still grown is alfalfa (15 percent) and sorghum (5 percent). In addition, three small pomegranate orchards (18 percent) are still maintained. The old hand-dug well of Bir al-Hadi was 54 metres deep and was used to irrigate mainly cereals. Sorghum (*dhurah*) was irrigated on half of the land (0.7 hectares) during the summer (*sayf*) and wheat (*burr*) on the remaining half during the winter season (*shita*). In the late 1970s the old well was deepened and equipped with a mechanical pump. By the mid-1980s pump volumes gradually decreased. A second pump was drilled to 200 metres but already by 1992 showed signs of diminishing volumes. A third pump, drilled during 1990, has been the cause of conflict between two families and is now out of

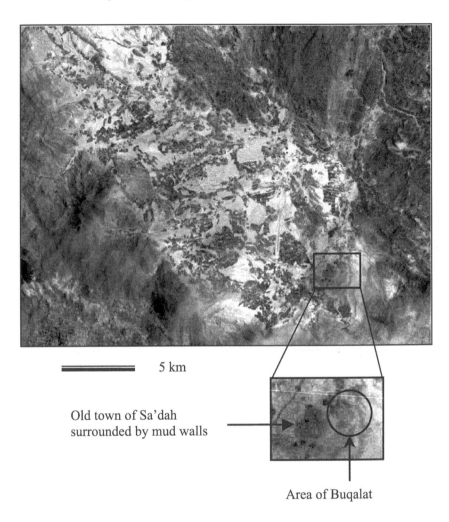

5 km

Old town of Sa'dah
surrounded by mud walls

Area of Buqalat

Figure 6.19   The area of Buqalat
   just east of the old mud-walled town of Sa'dah where land has been
   taken out of production as a result of water scarcity and high levels of
   salinity. Growing alfalfa, which tolerates salinity, has been one of the
   coping strategies. Given that Buqalat is within 15 minutes walk to the
   Sa'dah market and the high demand for fodder to feed the livestock of
   the town alfalfa is a lucrative cash crop.
   Date of satellite image: August 1998.

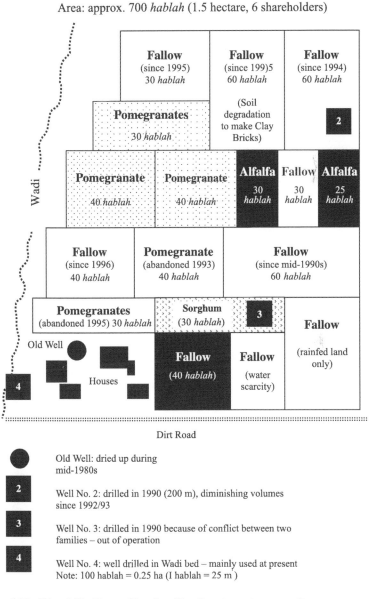

Area: approx. 700 *hablah* (1.5 hectare, 6 shareholders)

Fallow
(since 1995)
30 *hablah*

Fallow
(since 199)5
60 *hablah*

(Soil
degradation
to make Clay
Bricks)

Fallow
(since 1994)
60 *hablah*

Pomegranates

30 *hablah*

2

Wadi

Pomegranate

40 *hablah*

Pomegranate

40 *hablah*

Alfalfa
30 *hablah*

Fallow
30 *hablah*

Alfalfa
25 *hablah*

Fallow
(since 1996)
40 *hablah*

Pomegranate
(abandoned 1993)
40 *hablah*

Fallow
(since mid-1990s)
60 *hablah*

Pomegranates
(abandoned 1995) 30 *hablah*

Sorghum
(30 *hablah*)

3

Fallow

(rainfed land
only)

Old Well

Houses

Fallow
(40 *hablah*)

Fallow
(water
scarcity)

4

Dirt Road

Old Well: dried up during
mid-1980s

Well No. 2: drilled in 1990 (200 m), diminishing volumes
since 1992/93

Well No. 3: drilled in 1990 because of conflict between two
families – out of operation

Well No. 4: well drilled in Wadi bed – mainly used at present
Note: 100 hablah = 0.25 ha (I hablah = 25 m)

Figure 6.20   Bir al-Hadi *waqf* land: adjusting to water scarcity

operation. A fourth well has been drilled right at the edge of the wadi in the hope that seasonal spates will help replenish groundwater volumes in the immediate area.

The *waqf* land of Bir al-Hadi's 1.7 hectares is shared by six families. Demand for livestock fodder in Sa'dah town provides one family in Buqalat with some needed daily income. Water scarcity has forced them to gradually reduce the area they are able to irrigate, from 150 to 100 *hablah* (0.375 to 0.25 hectare) around the time of the Gulf Crisis in 1990/91 and from 100 *hablah* down to about 50 *hablah* (0.12 hectare) at the time of the Yemeni Civil war in 1994. At present just 40 *hablah* (0.1 hectare) of alfalfa guarantee them daily sales worth YR 300, nearly as much as the wage of an unskilled day labourer in the area during 1997. The family also owned pomegranate trees, which have been left to die when water scarcity forced them to make a choice. They own the share to only one day per week from the pump near the wadi and given the decreasing volumes it takes a whole day now to irrigate the 0.1 hectare area cropped with alfalfa. Given the family's poor circumstances they had obvious reasons to favour alfalfa over pomegranates. Alfalfa provides a secure daily income compared with the once-a-year return from pomegranates. Moreover, pump-sharing arrangements can complicate and delay necessary repairs in case of a breakdown. This poses a risk to fruit farmers who depend on secure supplies of irrigation water. Growing qat has not been an option because of *wabal*, a weed which 'has long stringy roots like our Cambridgeshire twitch grass' (Serjeant 1995 VII:72). While it is good for fodder (*alaf*) and reduces salinity of the soil, it weakens the crop and is difficult to control (ibid). A second family has started to turn the soil they rent from the Ministry of Religious Endowment (*wuzarat al-awqaf*) into clay bricks, which are sold at three Riyal a piece (**Figure 6.21**).

Adjacent to Bir al-Hadi is Bir Abu al-Layl where the situation is similar. The old well dried up during the early 1980s. A new well was drilled further east but operates with low output and much of the area is left uncultivated. **Figure 6.21** (bottom photo) shows parts of Bir Abu al-Layl area. In the foreground a plot can be seen where both a lack of irrigation water and the presence of *wabal* (weed) constrain agricultural production. The large house in the background is the home of an influential trader from the Sa'dah area. It provides a striking contrast to those who try to eke out a living from the small plots of Bir al-Hadi and Bir Abu al-Layl. However, it is said that the trader can be approached by farmers for financial help. Recently, he built a primary and secondary school near his home to meet the educational needs of Sa'dah's younger generation. As for the depth and number of wells that irrigate his land and supply for the domestic demand of over 100 people that reportedly live in the castle-like structure ... *Allahu a'lam* (God knows best).

Figure 6.21   (A) Turning the top soil into clay bricks
Coping strategies in areas suffering from water scarcity and high
levels of (top photo).
(B) Large house of a trader
Wabal weed (centre of bottom photo) presents an additional constraint
to agriculture in the area.

Resource Capture and 'Strong' Men

> Woe to you who add house to house and join field to field till no space is left and you live alone in the land (The Holy Bible Isaiah 5:8).

When coping strategies such as those described above run out, people with land come under pressure to sell. This facilitates resource capture further. In the water-stressed area of al-Dumayd this process has produced some intriguing outcomes. The photo in **Figure 6.22** shows a farm of 4,500 *hablah* (11.25 hectares) with an interesting tribal-political history of resource capture. In the mid-1980s, when groundwater levels were still at around 40 metres, a first well was drilled to 250 metres. Only two years later, by 1988, it started to dry up. A second well was drilled on the farm during the same year but volumes were not promising and it was left. When a third well was drilled during 1990 it became apparent that groundwater levels had dropped to between 80 and 100 metres. The owner, a local man, decided not to invest any more and abandoned the farm.

Figure 6.22   The farm of trader Aqlan
Aqlan was involved in a huge-scale car fraud business deal. Creditors seeking compensation have occupied the land. Given the water stressed condition of the al-Dumayd area farming on this land has been stopped and the pomegranate trees have withered and dried up.

Two years later, in 1992, the land was sold to the shaykh of Bani Mu'adh, although the farm is situated in the territory of the Walad Mas'ud. The shaykh then sold the farm to a trader from Kharif (Hashid). Incidentally, the man happens to own properties near the shaykh's hotel in Sa'dah town. The trader comes from the same place as Hashid's influential shaykh Mujahid Abu Shawarib, with whom the shaykh of Bani Mu'adh enjoys friendship. The Hashidi trader, however, sold the land to a certain Aqlan, a car dealer from Sinnatayn[15] (Hashid), who hit the local news with a big story of fraud during 1994, the year of the civil war. Aqlan had made deals with tribes from the east. They supplied the Hashid dealer with hundreds of new and expensive four-wheel drive vehicles, most of which had entered the country untaxed. In turn, Aqlan had promised to pay the owners a high price within 4–6 months of delivery. In the spring of 1994 these vehicles could be seen just off the road in a number of makeshift car parks in Aqlan's home territory, al-Sinnatayn.[16] Worth hundreds of millions the deal suddenly collapsed, Aqlan was imprisoned and the vehicles were confiscated. At the time many tribal roadblocks indicated the anger of those who lost money.

The family from Sanhan (Hashid) who now occupies Aqlan's farm in al-Dumayd is among those to whom Aqlan owes 'big' money. Trying to settle accounts they inquired about his land holdings in the Sa'dah basin. On finding out they simply moved north to become squatters on Aqlan's farm. By 1999 the family had not resumed agriculture on the land and photo in **Figure 6.22** shows that the former pomegranate trees have died. The case indicates the politicised nature of resource capture in the al-Dumayd area. The short visit to the farm and taking the photos, undoubtedly, raised many concerns and questions among the squatters.

Aqlan's farm is not an isolated incident of resource capture in the al-Dumayd area. Not far from the imprisoned trader's farm, one of Sa'dah's shaykhs has recently acquired 25,000 *hablah* (62 hectares) for YR 50 Mill. This amounts to YR 2000 per *hablah* and corresponds to the local price for land paid in 1999. Farmers reported that the shaykh had then sold most of the land with a profit of 300 percent to a very 'big' tribal actor who enjoys enormous political influence in national and regional politics. Buying at YR 6,000 per *hablah* the deal was said to have been around YR 130 million. In 1999 pomegranates had just been planted on the land reflecting the crop's present comparative advantage over citrus. Even though the Bani Mu'adh shaykh re-sold this large plot of land he keeps a number of stakes in the territory of Walad Mas'ud where he reportedly owns three smaller farms (2.5, 1.5, 5.0 hectares).

The area south and south east of Suq al-Talh marks a large runoff zone (*mahjar* pl. *mahajir*) which is shared by the tribes of Bani Mu'adh, Walad Mas'ud and al-Talh. In the 1980s claims to runoff and land rights led to tension between tribes and communities and privatisation was stopped largely to avoid government intervention. By 1999 some of al-Talh's influential traders and shaykhs had successfully negotiated a settlement for a long-standing dispute between al-Talh and the village of Al Abin. In return, they have been able to acquire a share of the disputed land. Increasingly, some of Sa'dah's shaykhs and traders are being accused of using their political and economic influence to settle and/or steer land disputes in their favour, often in return for more land. In some case, these actors are even said to stir up unresolved land issues so that they can propose mediation and subsequently acquire land in the process. In private conversation with informed individuals from the area it is not unsual to hear them say that *al-mashayikh hum sabab al-fitnah, al-mashayikh akbar sabab li-l-mashakil* (the shaykhs, they are the reason behind intrigue and the main reason behind the problems) or words to that effect.

Figure 6.23 shows part of the *mahjar* with the village of Al Abin in the background. The two darker lines running across the *mahjar* towards the village are *masqas* (low earthen dams). They had been bulldozed only in the mid-1990s, as a result of the recent settlement over runoff claims. These *masqas* mark the runoff zones and direct the runoff to the agricultural fields in the village.

Resisting Resource Capture

> No inheritance may pass from tribe to tribe
> (The Holy Bible, Numbers 36:9).

> If one of your countrymen becomes poor and sells some of his property, his nearest relative is to come and redeem what his countryman has sold[17]
> (The Holy Bible, Leviticus 25:25).

In addition to the dispute between al-Talh and Al Abin there is also an interesting dispute at the village level of Al Abin, which has prevented the privatisation of their grazing areas up to this point. Tensions evolve around the question of eligibility; who counts as part of Al Abin and therefore has the right to a part of the grazing land.

Al Abin is part of Walad Mas'ud (Sahar). During a prolonged period of drought in the 1980s a number of families from Hamdan al-Sham (Bakil) migrated from the east to Al Abin where they were allowed to buy land and to settle. These newcomers are now demanding an equal share to the village's grazing land when the time comes for it to be privatised. So far, the 'old' population of Al Abin has refused the people from Hamdan al-Sham their request arguing that they have no ancestral rights to the land. It is the perceived shrewdness and cunning of many Hamdan tribes that has led people from Sahar to label them as *awsakh al-nas* (the dregs of humanity). In terms of groundwater, disputes like the one described above have saved parts of the Sa'dah aquifer from exploitation because it stopped the privatisation of common land, which would have facilitated the spread of new groundwater-irrigated farms.

Some up-stream areas of the large grazing land shared by three tribes and many communities, as seen in **Figure 6.24** have escaped groundwater development simply because communities have not felt the need to split up the land for the time being.

Figure 6.23   Runoff zones for the hamlets of Al Abin

Figure 6.24   Runoff and grazing land to the south-west of Suq al-Talh

There are other factors that have prevented further resource capture. In several cases villages and communities have closed ranks to prevent their own skaykh from buying their land. As many of the local shaykhs have come to command considerable social and political power, individuals are usually reluctant to deny their requests to acquire land. But in a recent case a number of villagers resisted the wishes of their shaykh when it emerged that the shaykh acted as a broker for a rich businessman in the capital. Consensus to sell the land had already been reached between the shareholders when one man found out that the shaykh's intention was to sell it to someone from outside their tribe.[18] Based on the notion of *juwarah* (neighbourhood law), a tribal concept which stipulates that land must not be sold to 'outsiders' (Kopp 1981:136; Dresch 1993:82), the man convinced his fellow villagers to stop the land sale. Why should the Haves come and siphon off the water from the Have Nots' is the perception of many local people.[19]

A similar attempt by the same shaykh to buy common land from one of his village communities has also been unsuccessful. It is said that the shaykh offered YR seven million (US$56,000) to purchase an area of about ten hectares. However, the deal required the unanimous consensus of the 80 tribal shareholders. While over 70 people accepted the offer, the deal was blocked by a handful of individuals who did not think that their shaykh needed more land.[20]

There are exceptions to Sa'dah's shaykhs, many of whom engage in resource capture. In a rare precedent one shaykh has halted resource capture of the village land. Achieving consensus with his village community the shaykh has asked that no individual member sell part of his land to people from outside the community. His village is located close to the area of al-Dumayd which has experienced the disastrous consequences of groundwater mining. Moreover, the community has also learned the valuable lesson that selling land to people from outside their tribe complicates community water management.

Runoff Zones

Do not move an ancient boundary stone
(The Holy Bible, Proverbs 23:10).

The 1998 satellite image (see Figure 2.7) indicates a number of large runoff areas in the Sa'dah basin that have escaped groundwater mining. Locally, these areas are called 'white land' (*ard bayda*) reflecting the colour of sand and gravel. On the satellite image these areas also show up white in contrast to the groundwater irrigated areas reflected in red.

Runoff zones in areas where it is known that groundwater levels are falling rapidly are likely to remain undeveloped for irrigated agriculture. Farmers without additional income from non-agricultural sources can no longer afford expensive supply management options. Many of them now prefer sustainable rainwater harvesting agriculture over unsustainable and expensive groundwater irrigation options. For example, this trend has been observed in the water-stressed area of al-Dumayd during fieldwork in 1997.

Figure 6.25 shows a runoff zone and its adjacent agricultural fields being prepared for rainwater harvesting just south of al-Dumayd. Until the mid-1970s such rainwater harvesting characterised most farming activities in the Sa'dah basin. The bottom photo also provides a striking contrast between sustainable rainwater harvesting (foreground) and the unsustainable groundwater-irrigated farming (background, top right of photo).

A careful look at the runoff zone (centre of top photo) reveals the *masqas* (low earthen dams) that divide the larger zone conducting runoff to each individual plot. Unexpected rainfall occurred during the autumn of 1997, when the picture was taken, and a farmer is seen preparing the ground for sowing winter wheat (*burr*).

In contrast, **Figure 6.26** shows an area of yet undeveloped land down stream of the same *mahjar* and close to the main road that runs through the basin. This land is no longer attached to runoff rights and could be sold and/or developed for irrigated farming. However, given the precarious groundwater situation in the area it is unlikely to attract prospective buyers. As one informant from the area put it 'no one wants to buy land in al-Dumayd where the water is *ba'id* (deep), scarce and uncertain, and where new wells have to be drilled to 400 metres.' The picture shows land just off the main road where it sells for YR 5,000/*hablah* (US$ 16,000 per hectare) compared with 20 times that value, YR 100,000/*hablah* (US$ 320,000 per hectare), along the road. The plot marked by a high mud wall, although owned by one of Sa'dah's leading trading families, has remained undeveloped. A farming neighbour had intended to buy two plots adjacent to his farm for the future of his children. With little known groundwater, however, it made no sense to him and now the farmer is looking elsewhere to buy land where his children can try their own luck in farming.

Government/Institutions

'There is thus a rich agenda for reform at the sectoral level. The challenge as always in Yemen – will be effective implementation' (Ward 1997:10). Yemeni civil law is based on two sources: the Yemeni constitution and Islamic *shari'ah*. The constitution, in turn, stipulates that the *shari'ah* is to be the sole source of the law (Kohler 1999:141). With regard to Yemeni water law a number of sources have pointed out the contradiction between the constitution and the *shari'ah* (al-Eryani et al 1995:46; Haddash 1996:5; Yemen Times Feb 23 and March 2 1998). Whereas according to the Islamic *shari'ah* water can not be privately owned, the 1994 wording of the constitution (Article 7) states that:

Figure 6.25 Rainwater harvesting versus groundwater irrigation
(al-Dumayd area in autumn of 1998)
(A) The *masqas* (earthen dykes), which channel the runoff to separate plots, can clearly be seen above the agricultural land, in the centre of the picture (top photo).

Figure 6.26   An undeveloped plot of farmland
Land in the water-stressed area of al-Dumayd is likely to remain
undeveloped since people are aware of the fast-dropping groundwater
table.

All natural resources and sources of energy, whether above ground, underground, or in the
territorial waters.... are owned by the government, and it will ensure exploitation of such
resources for the common good.

The above quote allows for three important points to be made. First, Article 1366
of Yemen's Civil Law, which is based on the two sources above, appears to support
the *shari'ah* view when it states that 'water is originally res nullius for all *(mubah)*'
(Kohler 1999:141; Eryani al-, et al., 1995:46). Secondly, it might be argued that a
resource defined as 'res nullius' could be understood as a *Common Property
Resource (res communis)*. In this sense ownership of the resource (water, for
example) by the government might be viewed as 'stewardship' to 'ensure
exploitation of such resources for the common good'. However, in politicised
environments, such as the Sa'dah basin, and indeed many Yemeni regions, this
interpretation is unlikely to gain currency as long as the government is perceived as
lacking credibility and legitimacy. Thirdly, with regard to groundwater resources in
particular, the question of ownership and/or stewardship by the state runs contrary to
established *shari'ah* interpretations. According to Islamic law water from a res

nullius source contained in a well drilled on privately owned land becomes the property of the well owner. (Kohler 1999:142, quoting Eryani al- et al. 1995:47).

Legal pluralism, the contradiction between constitution and *shari'ah*, described above, presents only one set of challenges for water resource management in Yemen. Efforts to address water issues have also been characterised by the overlapping and often conflicting interests of various ministries and institutions, at least, until the mid-1990s, when one central institution was formed to oversee all aspects of water allocation and management (Kohler 1999:142f).

During the 1980s a number of government ministries shared responsibility for the country's water resources: the Ministry of Agriculture and Water Resources (MAWR), the Ministry of Oil and Mineral Resources (MOMR) and the Ministry of Electricity and Water Resources (MEW). Kohler (1999:143) points out the conflict of interests this created. MAWR tried to safeguard the vested interests of farmers, who represent the heaviest user consuming over 90 percent of the country's water resources. In contrast, MEW lobbied to meet the rising demand for domestic water in towns and large cities. In between these two institutions MOMR was concerned with hydrogeological surveys and the collecting of data to ensure optimal and sustainable uses of water resources.

In 1981 the High Water Council (HWC) was formed to co-ordinate between the various institutions and to provide guidelines for the formulation of a national water policy. Its failure to reach these objectives is attributed to the fact that the council was chaired by the Minister of the MEW, a politician who also represented the interests of one of the institutions in charge of water (Kohler 1999:142).

The situation improved in 1986. The HWC was formally placed under the Prime Minister's office and a 'Technical Secretariat' was also added. The draft proposal, in 1990, for a new water law also benefited from the support of international organisations. The HWC was formally supported by the UNDP. Moreover, in partnership with the Dutch Institute for hydrological studies (TNO) MOMR's General Department of Hydrology (GDH) prepared a number of studies and surveys, which guided and influenced the proposal. The GDH has argued for a leading role in the water sector because, unlike MAWR and MEW, it sees itself as a non-user of the resource and, therefore, with no vested interests. However, the leading role of the GDH within the HWC has often resulted in detailed hydrogeological surveys at the expense of studies to shed light on the socio-economic and socio-political complexities that influence the allocation and management of water (Kohler 1999:143). In retrospect to the GDH's missions in the Sa'dah basin this was recognised by TNO's principal hydrologist when he writes that '[O]viously, there is a need for a model that describes and explains the allocation of land and water resources in the area of crops, as a function of socio-economic factors' (Van der Gun 1987:8). Moreover, as shown in Chapter 4, hydrogeological surveys conducted in politicised environments are likely to be viewed with suspicion by stakeholders and can trigger responses that adversely impact on water demand management.

Mutual suspicion and competition for power and influence between institutions slowed down the work of the HWC. MAWR provides one interesting example when, in 1991, it came up with its own proposal that had been drafted independent from that of the HWC.

The political and economic instability in the wake of the 1990/91 Gulf Crisis, which culminated in the 1994 military confrontation, delayed progress on water policy and law even further. However, March 1995 saw the creation of the National Water Resources Authority (NWRA) to supersede the HWC which was dissolved. As a new institution NWRA is empowered with a new concept, different structures and increased financial and political capacity. Its increased political authority is evident from articles 6 and 7 of its manifesto.

> The Authority is the sole governmental agency in the Republic responsible for the formulation of water resources policies and development strategies and the study, planning and management of water resources at national level...
> The Authority has all the powers and jurisdiction necessary to achieve its objectives...
> (NWRA 1996:11, as quoted in Kohler 1999:144)

NWRA understands that in many Yemeni regions the last phrase 'has all the power and jurisdiction to achieve its objectives' carries little weight. For this reasons the new approach recognises the importance to involve communities and stakeholders at all levels of planning, discussion and implementation (World Bank 1997:18).

NWRA's responsibilities include three main areas. First, the 'policies and programming sector' has the challenging task to formulate and draft a new water law. Furthermore it oversees the co-operation between national and international water-sector institutions. Secondly, the 'studies and information sector' deals with the collection and analysis of data and information specific to the country's water resources. Lastly, NWRA's 'implementation and monitoring sector' is to ensure implementation of new water policy. It is this last sector which has to foster the close co-operation and participation of stakeholders which is vital for the success of NWRA.

Table 6.1 presents an overview of the Yemeni water legislation. A comparison between these water laws indicates a number of trends and developments as indeed observed by Kohler (1999:152). First, it will be noted that *shari'ah* law bears directly on customary law, constitution and civil law but does not figure formally in the proposals made by HWC, MAWR and NWRA. This is problematic because the contradiction inherent in the constitution with respect to water ownership (state-owned versus *mubah* 'res nullius') will remain a contentious issue unless new Islamic interpretations can be worked out.

Secondly, with respect to ownership of water the government has shifted its position. In the NWRA wording the country's water resources are no longer owned by the state but are declared 'public property'. The state becomes the steward rather than the legal owner of the resource with the responsibility to develop the resource for the common good. However, the issue of legal pluralism remains unresolved.

Table 6.1 Comparative water law in Yemen (Kohler 1999:152)

<table>
<tr><th></th><th colspan="7">Comparative Water Law in Yemen</th></tr>
<tr><th></th><th>Customary Rights</th><th>Constitution 1991</th><th>Civil Law 1994</th><th>Proposal HWC 1990</th><th>Proposal MAWR 1992</th><th>Proposal NWRA 1996</th></tr>
<tr><td>*Wording*</td><td>Unwritten law</td><td>1 out of 131 articles</td><td>30 out of 1399 articles (land & water rights)</td><td>Water law with 44 articles and 'by-law'</td><td>Law with 50 articles</td><td>Law with 99 articles and policy</td></tr>
<tr><td>*Reference to Shari'ah*</td><td>Strong</td><td>Yes but contradicting</td><td>Strong</td><td>No</td><td>No</td><td>No</td></tr>
<tr><td>*Ownership*</td><td>*rescommunis* Private ownership for contained water</td><td>State</td><td>resnullius, water belongs to all</td><td>State</td><td>State</td><td>'Public property' managed by state</td></tr>
<tr><td>*Control*</td><td>Landowner</td><td>State</td><td>Not specified</td><td>National Water Authority</td><td>Water authority in MAWR</td><td>NWRA</td></tr>
<tr><td>*Management*</td><td>Decentralised</td><td>Not specified</td><td>Not specified</td><td>Centralised</td><td>Limited decen-tralisation</td><td>Decentralised management explicitly demanded</td></tr>
<tr><td>*Priority of use*</td><td>1. Drinking 2. Irrigation</td><td>Priority for 'common good'</td><td>Not specified</td><td>1. Domestic 2. Economic criteria</td><td>Drinking</td><td>Economic criteria 1. Domestic 2. Industry 3. Agriculture</td></tr>
<tr><td>*Measures to control over-exploitation*</td><td>No direct control</td><td>Not specified</td><td>Not specified</td><td>Strongly controlled by law</td><td>Less strongly controlled by law</td><td>Decentralised management and permits</td></tr>
</table>

Shari'ah is the source of law in Yemen according to which groundwater contained in a well on privately owned land is the property of the landowner. The state cannot claim management of these water resources. Also, the contradiction with the constitution remains, where the state is granted ownership of all natural resources.

Thirdly, it is significant that in contrast to the two earlier water law proposals the NWRA Draft envisages a decentralised structure of management. This development is much more in line with traditional structures and customary law and takes into account the country's socio-political realities. Likewise, the NWRA draft delegates more power to local communities to control groundwater abstraction and drilling of wells.

Fourthly, all three water laws agree that water for domestic consumption should be given sector-priority.

As Kohler notes, in the new draft water law the state justifies control over, and a quasi ownership of, the country's water resources by arguing that it will ensure the use of such resources for the common good (1999:153f). And indeed, in one instance the Islamic principle of 'no harm' (*la darar wa la dirar*) provides justification for the state to intervene in customary and established water rights.

> when conflict arises between harm (to one party) and benefits (to another) then 'action aiming at averting spoilage takes precedence over availing benefits' (*dar' al-mufasid muqaddam ala jalb al-musalih and al-ta'arud*). That is, averting the spoilage arising from water shortages (dessertion [sic] of the city, spread of diseases, etc.) takes precedence over availing the benefits of irrigation (al-Eryani, et al. 1995:66).

While the above quote appears to provide justification for the state for present and future water transfers from rural areas to urban centres the *shari'ah* nevertheless limits application of this principle to the abstraction of water for drinking and for ablution before ritual prayers (Kohler 1999:154). At the same time al-Eryani, et al. 1995:67) make it clear that 'desperation of one party doesn't annul the right of the other' and that 'current water users [who may be adversely affected by the project] have the right to, and should receive fair compensation.' And this is precisely where one of the problems lies. Past experience has convinced rural communities in Yemen that the government has not honoured compensation agreements. In other cases promised benefits have been hijacked by individuals, often tribal leaders from the very communities that were meant to be compensated.[21]

Given the limitation of state control and influence in many tribal areas, especially in the north and east of the country, and the prevailing perception that the government apparatus is corrupt and lacks credit as an honourable partner it is doubtful whether and to what extend NWRA can fulfil its mandate to legislate, monitor and manage the country's scarce water resources. As long as the majority of people in the Sa'dah region feel excluded from the benefits of progress and development promised to them by the state any water 'owned' will be perceived as a precious and highly valued private asset and as a resource which continues to guarantee a certain amount of freedom, autonomy and continuity.

By June 2000 more changes to the earlier NWRA draft were discussed. The question of what sector should be given priority use remained a contentious issue. While there was agreement that the domestic sector be given first priority there were heated debates whether industry/tourism or agriculture should take second place. In addition, political power shifted away from NWRA, who appeared to become relegated from an executing to a technical authority. In order to combat the looming water crisis three main strategies were considered; water harvesting; saving water and wastewater reuse. One specific measure resulted from years of debates; pumps may no longer be imported. In the meantime, the qat issue was addressed at a high-level conference.

Developments took a sudden turn when the long-awaited water law was finally ratified by the parliament on July 22, 2002. The final version is an Arabic document, which has been translated into English by the World Bank. Ownership of water and water rights, contentious issues in previous drafts, are now defined in Article 4 as follows:

> Water is a right that is accessible to all and does not become privately owned except by means of transport, acquisition or any other related methods an (sic) it is optimal and is secured by its similitude.[22]

Moreover, the new water law gives NWRA the sole responsibility and the mandate for water management within the Republic of Yemen.

Conclusions and Summary

Firstly, this chapter has shown that a politicised environment, such the Sa'dah basin, allows for a range of coping strategies with respect to water scarcity. However, while some of these coping strategies suggest the presence of adaptive capacity others indicate that a politicised environment diminishes the adaptive capacity of actors and social entities.

Secondly, the case studies have indicated the potential and constraints at work in politicised environments, such as the study area, to implement technical (productive) and economic (allocative) efficiency. It was found that the ongoing process of resource capture by influential actors, especially certain local tribal leaders and traders serves to politicise water thus decreasing the level of legitimacy and by introducing elements of mistrust which undermine efforts to introduce water demand management strategies.

Thirdly, there is evidence, however, that forms of adaptive capacity exist. Some of the case studies have indicated that perceptions of the value of water are changing and that communities and individuals are starting to learn lessons from the mistakes of others. Moreover, in some cases resource capture is being resisted and people are starting to take a longer-term view with regards to their water resources. In this context it has been shown that in a number of cases rainwater harvesting, where it is still possible, is preferred over groundwater-irrigated agriculture.

Finally, this chapter has looked at the changing dynamics of governmental water institutions to draft a modern water law for the country. Significantly, the recent 1996 NWRA proposal recognises the importance of decentralised management and the crucial role of local institutions and community participation to ensure implementation of the new water law.

The next and final chapter will review some of the developments, themes and lessons discussed in Chapters 1–6. It will then look at the possible role of religion and poetry as a way to increase awareness and change perceptions of the value of water, and to address the water problem using media forms that are culturally relevant and socially acceptable, or to use a phrase that was recently coined 'politically feasible ameliorative alternatives'.[23]

Notes

[1]   See also Milroy A., (www. al-bab.com./yemen/env/arid.htm) 'Believing these newly-accessed groundwater supplies to be *al-bahr* – the sea – and thus limitless, farmers have increasingly relied in the past decade on these pumped aquifer water for agriculture, drinking and livestock.' (page 4f)
[2]   Air-lift pumps consist of a drop pipe placed in a well with its lower end submerged. An air pipe delivers air at the bottom of the drop pipe and forms a mixture of air and water which is lighter than the column of solid water in the well; consequently, the mixture rises above the surrounding water.
[3]   This information was obtained through conversation with Alison Brown, horticulturist, SOAS and Wye College, London, October 1997.
[4]   Handley (1996:7) reports similar cropping patterns for cereals in the Amran valley.
[5]   For qat Morris (1986:128f) notes: 'Since the late 1970s all qaat entering markets controlled by the government has been subject to a levy equivalent to approximately ten per cent of its retail value.'
[6]   Amran limestone aquifer units make up much of Al Ammar's territory. 'These units constitute poor aquifers and from the point of view of regional groundwater flow could even be termed 'aquitards'. Groundwater storage and flow seem to be predominantly related to fissures and cracks, and consequently, hydraulic characteristics may vary considerably from one location to another.' (Van der Gun 1985:26)
[7]   Qat comes in all kinds of qualities and prices but this figure appears very high. It is six times as much as reported by a qat farmer in the Sa'dah basin who reported cross benefits of YR200,000 for 130 *hablah* qat (YR1.2 million per hectare). In a 1993 crop budget study (World Bank, 1993:46 Agricultural strategy) the gross output per hectare for qat was calculated at YR 720,000.
[8]   Dresch (1993:27) mentions Wadi Madhab 'where there seems always to be flowing water: at several places it is flanked by reeds and frequented by wading birds, but the river makes a very thin ribbon through the arid landscape.'

[9] The author knows of one landless family from Sa'dah town which helps with the annual sorghum harvest in Wadi Madhab in return for fodder (*alaf*) for their sheep.

[10] Frost puts limits to qat production in other highland basins in Yemen. In the Amran basin, for example, qat is grown in protected narrow valleys. Moreover, it is sometimes intercropped with winter barley or planted alongside fruit trees for protection against frost (Handley 1996:7).

[11] On the different use of daggers and rifles see Dresch (1993:39; 86).

[12] Maktari (1971:36f) quotes an interesting *fatwa* (Islamic legal opinion) by Ibn Hajar concerned with this problem: 'he who wants to water his land in accordance with custom may do so, and he should not be deterred even if he inflicts damage on his neighbour. He is not as such responsible for loss [of property] resulting from his action.' Maktari extends Ibn Hajar's interpretation to include the drilling of a well: 'If a person digs a well within his property in the accustomed manner ...and this causes [his neighbour's] not to function, the digger of the well is not responsible...'

[13] The Spanish island of Ibiza provides a case study in the European context. In the early 1990s farmers from central Ibiza sold soil to new elites and migrants building villas along the island's cost. Coming from northern European countries with much wetter climates, many of these new residents have sought to create the same lush gardens (which they enjoyed back home) in this more water-scarce southern environment. In response to demand for fertile garden soil local farmers from central Ibiza have sold the soil of their own farms. Economic factors for selling the soil included water scarcity, the availability of cheap fruit and vegetable imports through the European market and the need for cash (Private communication in San'a 18 June 2000 with Prof. Miquel Barcelo, University of Barcelona who himself is a citizen of Ibiza).

[14] Dresch (1993:13) goes as far as to say that 'farming is reckoned perfectly honourable, and a man who did not farm would in most areas be reckoned no tribesman.'

[15] Sinnatayn is associated with trade and transport much of which passed through the Sa'dah basin. This process of smuggle and trade facilitated all kinds of links with actors in the Sa'dah basin. For the early 1980s Dresch (1993:308) notes that the village of Sinnatayn 'was credited with by some with as many as three hundred trucks; four- or six-wheel Mercedes chassis fitted with high box bodies for dry goods or with tanks to transport petrol.'

[16] Travelling the road through Sinnatayn frequently before and during the 1994 Civil War this author personally witnessed the build-up of Aqlan's car parks in the area.

[17] According to the laws in the Pentateuch '...individuals owned land only as representatives of their tribe. No one was allowed to transfer land outside the tribe (Numbers 36:5ff), nor to sell it to anyone in perpetuity. Every fifty years, in the Year of Jubilee, all land was to revert to its original owner.' (Stott, 1984:114f)

[18] This is confirmed by Kopp (1981:136) who states that even after the signing of a land deal to an 'outsider' any tribal member has the right to intervene demanding that the deal be annulled.

[19] The concept of 'juwarah' has recently also found application in the small German town of Grävenwiesbach (Hesse). Its location within the commuter belt of Germany's financial capital, Frankfurt, has attracted land speculators and investors in real estate from outside the community. The community council decided to ban 'outsiders' from buying land for building. Residents must have lived 17 years in the community before they are able to sell a building plot. Moreover, residents who purchased a plot must commence building a house within two years of purchasing the land. These regulations were triggered by developments in the near-by town of Usingen, where the price for a square metre has sky-rocketed reaching DM 1,000 compared with DM 212 per square metre in Grävenwiesbach. The town council wants to

ensure that the people of its own community are able to afford land for building and that there will be plots for its future generation.

[20] Morris (1986:131) makes the point that '*maal* (wealth) is synonymous with land. Land is the most common cause of dispute and great resources may be deployed to maintain claims to it.'

[21] Chris Handley, private communication.

[22] Arabic *wa huwa mithli yadman bimithlihi* 'If [the resource] is guaranteed to be replaced in kind' (author's translation).

[23] Personal conversation with J.A. Allan, May 2000.

7 Conclusions

kullu waqt lahu hall
Each time has its own solution
(Sa'dah famer, 1999)

The initial hypotheses that Sa'dah's tribal society has the capacity to address the issue of groundwater mining co-operatively can not be verified. First; it has became apparent that the expansion of groundwater irrigated agriculture from the late 1970s to the mid-1980s was exacerbated by socio-economic and political forces specific to Sa'dah's politicised environment. Secondly, current water use patterns seem to reflect not only economic but also cultural and political values associated with particular crops. And lastly, observations suggest that the region's political economy and its politicised environment explain potential and constraints for adaptive responses of communities and individuals.

Key Issues and Events

There are distinct phases to the development of groundwater in the Sa'dah Basin of Yemen. Prior to the early 1970s groundwater-irrigated agriculture was not possible because most of the land was communally owned and managed.

In 1972, a turning-point decision was made which changed this customary legal system. This decision was the result of arbitration by a Zaydi scholar who intervened in an inter-tribal feud. This scholar proposed that a community should give up half of its grazing land to those owning rights to the runoff from that land. This arbitration consequently altered the control and balance of power with regard to land and water resources.

The result of this was an immediate privatisation of common land. Thus started the quest for access to water in the alluvial aquifer. The Sa'dah hydraulic mission was born, manifesting itself as a relentless quest for access to the only source of available water – the alluvial aquifer.

It is argued that privatisation of common property resources helps to prevent a 'tragedy of the commons' (Johnston 1994:639; Ostrom 1990:12).[1] In this case study it appears ironic, perhaps, that the privatisation of tribal common land resulted in a 'tragedy of the commons' scenario for the basin's groundwater resources. The gap between prevailing religious beliefs and interpretations, which are based on pre-modern technologies and the possibilities afforded by modern pumps provide the answer. In the pre-technological era groundwater was a 'restricted access resource' since its abstraction was restricted by animal power and shallow wells. In addition, the 'restricted access' character of communal tribal land, in turn, imposed 'restricted access' on groundwater resources. No wells could be drilled where runoff rights were attached to communal land. It was the privatisation of communal lands from the mid-1970s on that facilitated a change of status for groundwater. Through land privatisation groundwater became an 'open access resource', precisely because it now gave a private landowner the right to abstract whatever volumes could be 'contained' in his well, as stipulated by Islamic legal interpretations. It is this shift from a 'restricted access' to an 'open access resource', which is gradually leading to a 'tragedy of the commons'.

Belief

Furthermore, the impact of this 'tragedy of the commons' scenario is accepted by a large majority of farmers on the grounds of two Qur'anic verses. These are:

> And we raise some of them above others in ranks,
> So that they may command works from others.
> But the mercy of thy Lord is better that the (wealth) which they amass.
> (*Al-Zukhruf,* Surat 43:32, Translation by Yousuf Ali)
>
> *wa Allah faddala ba'dakum ala ba'dan fi al-rizq*
> With respect to sustenance Allah has favoured some over others
> (*Al-Nahl,* Surat 16:71, translation by Lichtenthäler)

Consultation with a number of scholars confirmed the established Islamic belief that land owners have the right to utilise and abstract groundwater on their own property, as long as they do not waste the God-given resource. When the environmental consequences were repeatedly pointed out to these scholars, they usually replied that 'the state of people is all ordered by God'. Even when the details of the consequences of the four to six metre per year drop in groundwater levels were explicitly spelled out, these two Qur'anic verses were quoted in order to indicate that notions of equality in this life are not in accordance with prevailing normative systems.

The effect of this normative system is twofold when seen against the background of water resources management. Firstly, it serves to explain why some people have the capacity to abstract and mobilise more water than others do. This explains the amazing absence of tension and conflict over groundwater resources where land rights are firmly established in terms of the turning point 1972 arbitration. Secondly, it provides the basis for the legitimisation of resource capture, which is becoming evident as the water levels are dropping. The dynamics of this process are fuelled by the fact that as the water table falls, deeper wells need to be sunk and longer pipelines need to be inserted in them. This takes money, which for the poor is mobilised by selling land. In addition to this, the deeper water requires more energy to lift to the surface, so larger engines and increased operating costs are placing further pressure on marginal farmers. The richer farmers are able to buy up land cheaply and thereby entrench their positions of power.

Politics

From the mid-1970s to the late 1980s, the traders of Sa'dah were the principal importers of pumps and irrigation equipment. In addition to this, the same families often own the drilling rigs that are needed to deepen existing wells as the water table falls, and drill new ones. Thus, the expansion and development of irrigated agriculture has directly benefited a number of trading families and has helped them to achieve a significant degree of economic power. All traders own large farms that produce citrus and qat. Significantly, it was the traders who were the first to introduce measures to achieve productive efficiency, by experimenting with drip irrigation technology. The motives for this form of WDM were thus self-interest, rather than a sense of trying to conserve water as part of a sustainable development programme. Figure 7.1 shows the increase of irrigated agriculture in the Sa'dah basin up to 1998 as evident from a satelite image (August 1998). Comparing this image with a generalised version of agricultural land based on airphoto interpretation (Danikh and Van der Gun 1985: Fig 11) it becomes apparent that irrigated area has doubled, from 50.7 km^2 in 1980 to 99.9 km^2 in 1998. The satellite image, which was obtained only in 1999 when fieldwork had been completed, also confirms the extent of resource capture in the areas discussed in the previous chapters. Yursim (just east of the Sa'dah airstrip) and al-Dumayd (to the north-east) provide striking examples. One of the consequences of this resource capture is the introduction of feelings of mistrust into the community. This is particularly evident when a dispute arises needing some form of arbitration. Shaykhs and traders are regarded with suspicion, as they are perceived to have vested interests and therefore being incapable of impartiality.

The WDM strategies formulated by NWRA (National Water Resources Authority) are unlikely to be implemented effectively as there will be such an element of structural scarcity in existence that the efforts may be perceived by the ecologically marginalized people to be only to the benefit of the rich and powerful.

Many small farmers and especially those negatively affected by the changing economic and environmental realities wish the government to take a more pro-active role in groundwater management. But for any intervention to be successful a lot of trust has to be built first. The problem is that many of those representing the government, and those associated with it, own large farms themselves, which in turn reduces their credibility.

Government bodies concerned with water management in the Sa'dah basin must seek to restrain these actors from resource capture and so demonstrate that they can

5 km

Figure 7.1   Increase in irrigated area up to 1998.
Date of satellite image: August 1998.

not act above the law. A genuine commitment to apply its own rules to those in its service will help rebuild trust and lead to co-operation.

The Government has an important role to play in facilitating and supporting local water management. It will effectively do so if it can reassure the tribal communities that its intentions and its plans for water management are in the best long-term interest of their own communities. Unless corrected, resource capture serves to de-legitimise WDM strategies (Turton and Ohlsson 1999:11f).

Culture in Context: The Need for further Research

You must teach him four things,
– the dictates of Islam,
– how to shoot a gun,
– how to dance,
– how to compose poetry

A tribesman from Khawlan al-Tiyal on what it means to be a *qabili*
(Caton 1990:26)

The following points to the existence of social adaptive capacity. It also suggests these windows of opportunity can be explored further to develop some appropriate and highly innovative WDM strategies using traditional value systems and poetry as a vehicle for transmitting information.

A visit to the study area in 1999 revealed some interesting changes to cultural patterns of some of the tribes. Evidence was gathered that on several occasions a tribal community acted collectively in an attempt to safeguard their groundwater resources. This tentatively suggests that notions of sustainability are now starting to take root.

The Role of Religion

Many *sayyid*s are the doctors of religion with the Sa'dah basin being the centre of Zaydi scholarship. While many *sayyid*s became marginalized during the 1962 revolution, their services as teachers, preachers, arbitrators and judges have increasingly become sought after (Haykel, 1995:20f). Unlike the tribesmen, they usually own very little land and are therefore not perceived as acting out of self-interest when they are called upon to address a water-related issue. It was a Zaydi scholar who acted in the turning-point 1972 mediation vis-à-vis surface water and land rights. In the recent past a younger group of Zaydi scholars appear to have emerged. They are trying to reconcile Zaydi principles to the requirements of modern life. Significantly, they do not represent one specific tribal grouping and they have no vested interest in agriculture. They can thus be considered to have a high degree of impartiality and legitimacy.

In order to determine the impact of the Zaydi resurgence,[2] a number of scholars were visited. All of the scholars who were contacted were asked about the Quranic principles impacting on groundwater abstraction as noted above. All reconfirmed the established Islamic belief that land owners have the right to utilise and abstract groundwater on their own property as long as they did not waste this God-given resource. However, one very significant change in the Islamic interpretation was noted. An eminent Zaydi scholar who serves a tribal community that is particularly hard-hit by the falling groundwater table, quoted the following verse from the Qu'ran:

Say: 'See ye?- If your stream be some morning lost (In the underground earth),
Who then can supply you with clear-flowing water?' (Surat 67:30 *Al-Mulk*)

Surat Al-Mulk (dominion), from which this verse is taken, emphasises the fact that Allah in his goodness and power has ordered and arranged the world. If life-sustaining water suddenly disappears one morning, who can bring it back? The Zaydi scholar understood this to mean that farming communities do in fact have a collective responsibility to make sure that Sa'dah's common groundwater resources must be used in such a way as to avoid that the 'stream be some morning lost'. His interpretation stressed man's responsibility for managing his environment in a sustainable way. He also challenged the widely shared notion that God will readily reverse the damage done by decades of over abstraction. The Islamic principle of *maslahah ammah* could be given greater emphasis as it clearly recognises that the interests and welfare of the wider community have priority over and above individual rights and benefits, even if these are lawful (*al-maslahah al-ammah muqaddimah ala al-maslahah al-khassah*). The *shari'ah* has to be applied wherever the general interests lie (*haythuma kanat al-maslahah fathama shar Allah*). Based on the notion of 'no harm' it appears as if the concept of *maslahah ammah* could be explored to help regulate groundwater abstraction and well drilling.

Therefore, just as in the 1970s when religious scholars were brought in to solve the impasse over land and water rights, leading to the turning-point 1972 arbitration that led to land being sold for water rights, the same religious scholars can be used to introduce WDM strategies. This is becoming apparent in the increased levels of awareness among farmers that groundwater management may indeed serve the long-term interests (*maslahah ammah*) and result in sustainable development practices being introduced.

The problem of communicating the underlying hydrological conditions is a major challenge. The need to change water management practice requires a preliminary shift in awareness. Outsider professionals have little impact on local perceptions. Insiders who are respected are more likely to be influential.

The possibility of gaining the co-operation and help of Islamic preachers and teachers to help address problems should be considered. That this can be done effectively through religious sermons has been shown in another Middle Eastern country (IDRC 1998). Relevant knowledge about water issues is shared and discussed with Muslim preachers, who then incorporate the information into their religious sermons. Furthermore, Islamic scholars should be encouraged to explore ways in which the Islamic principle of *maslahah ammah* (public interest) can be applied in an attempt to restrict groundwater mining. Ultimately, ownership of any kind in Islam belongs to God alone. Man is only God's steward entrusted with the responsibility to ensure that God's creation is managed well and sustainably for future generations (Kohler 1999:166).

The Role of Tribal Poetry

Evidence uncovered during the period of fieldwork suggests that poetry could play a supportive role in shaping perceptions vis-à-vis principles of groundwater sustainability and water demand management.

Yemen has a long history of employing poetry for persuasion and in conflict mediation as shown by a detailed study carried out in *Khawlan al-Tiyal*, a tribal area south-east of San'a (Caton, S. 1990). Moreover, within the context of contemporary society, tribal poetry has been used effectively to criticise party politics, new elites and new power centres (Dresch 1995:5, 1995:417ff, Caton 1990:48). A brief appraisal of locally produced poetry in the Sa'dah area suggests that this literary genre could be utilised to increase community awareness and co-operation to bring about changes of perception vis-à-vis the sustainable and equitable use of the shared groundwater resources.

The Sa'dah region is renowned for its poetic conventions. Tribal poetry (*al-sha'r al-sha'bi*) has played an important role in peacefully solving some of the long-standing tribal conflicts in the area. For example, the conflict between the tribes of Bani Uwayr (Khawlan b. Amir) and Harf Sufyan (Bakil) during the 1980s went on for many years and took 20 lives, including the lives of two women, before it was eventually settled. Poetry played a crucial role in resolving the blood feud and the poetic words (*zamil*) are alive in the memories of tribesmen and city dwellers alike.

In order to determine whether poetry is a potential medium for communicating the need for WDM in achieving natural resource reconstruction via sustainable development practices, one of the foremost poets from Sa'dah was visited during a field visit. Hasan is a *sayyid*, claiming descent from the Prophet, and comes from the Khawlan area to the west of the Sa'dah basin. Over the recent past, Hasan has regularly written poetry for cultural and political occasions. At times his work has been commissioned by the Government for important political events such as the anniversary of the September Revolution. Other poetry he produced relates to the tribal and social context of his native Sa'dah region. In a discussion with the poet, a number of local men from various backgrounds were present. After noting the problems caused by excessive groundwater abstraction, all of the men present confirmed with some enthusiasm the role that poetry could play in changing this. Hasan enthusiastically volunteered for the task of composing a poem addressing issues of equity, sustainable abstraction of groundwater, co-operation over shared resources, irrigation efficiency (intra-sectoral allocative efficiency) and water savings. Significantly, Sa'dah's Cultural Centre (*al-markaz al-thaqafi*) has also expressed interest in Hasan's composition on the themes of environmental resource use and agricultural issues. They could provide valuable support in disseminating the poetry via the media and through the production of audiotapes. This poetry could also form part of cultural events that are held at the Centre, some of which have been televised in the past.

It was also confirmed that each tribal area has its own acknowledged poet who composes on behalf of their respective tribe and in response to specific issues. Tentative conclusions can thus be drawn that there is ample scope for the investigation of the potential use of poetry as a culturally accepted way of communicating information regarding groundwater abstraction in Yemen. Issues such as allocative efficiency, the dangers of resource capture and the benefits of sustainable development could be incorporated into a culturally acceptable WDM strategy. Such initiatives will generate and strengthen social adaptive capacity. However, it is proposed that further research on the religious and poetic theme be conducted in order to utilise this convention.

Lessons and Conclusions

Firstly, the Sa'dah basin is an example of a social entity that is facing a first-order scarcity of a fundamental natural resource – water. The only water available is that which is found in alluvial aquifers lying under the basin. These aquifers are being over-abstracted to such a degree that their levels are dropping at an alarming rate. It is safe to conclude that current abstraction levels are unsustainable, and in the absence of any other alternatives, the area will face a crippling water shortage in the next decade or two.

Secondly, the spatial distribution of irrigated agriculture in Sa'dah's politicised environment is a function of tribal-political differentiation as much as (or more than) environmental factors. The unsustainable mining of the Sa'dah basin's groundwater resources from the mid-1970s until the mid-1980s, can be explained as the outcome of unequal power relations, political interests, and the changing ability of actors to control or resist other actors.

Thirdly, the Sa'dah study provides excellent evidence of the complex relationship that value has as an aspect of the concept of return to water. Allocation and management of Sa'dah's water resources are strongly influenced by tribal-political perceptions of interests and power and by socio-economic and socio-political values. In this context economic rationality differs fundamentally from cultural rationality and political rationality. The latter is included because political rationality within a localised context relates to elements of prestige and power. Therefore, crop type, farm size and number of wells are visible symbols of wealth and political expression of power. Such notions militate against efforts to achieve economic (allocative) or technical (productive) efficiency. Where allocatively efficient methods are in place it is at the level of the production unit and is thus within the farm, involving a shift from one crop to another. The region's politicised environment, which inhibits the development of an industrial sector, explains the lack of inter-sectoral allocative efficiency. Ultimately, the lack of possible alternatives is a major constraint to achieving effective natural resource reconstruction through WDM strategies.

Fourthly, the Sa'dah study confirms the assumption of political ecologists who argue that social and economic cost of environmental degradation are distributed unequally among the different actors. There is evidence of ecological marginalization. This has been driven to a certain extent by the rapidly changing normative basis of society resulting from the 1972 turning-point arbitration on land and water rights. The rapidly falling levels of the water table have further exacerbated this, with poorer farmers simply being unable to afford the increased cost of well deepening, re-equipping and higher operating costs.

Fifth, the social, economic and political resourcefulness of individuals and communities has become evident. It was shown that politicised environments create and help sustain coping strategies not available to political economies controlled and regulated by a central government. At the same time such environments do not necessarily foster adaptive capacity, indeed they are likely to undermine it, especially where resource capture is occurring as evident from the Sa'dah study. Up to the time of writing the resource capture of groundwater is justified and tolerated in terms of traditional customary and religious perceptions. If this is allowed to continue, it is likely to undermine subsequent efforts at introducing WDM strategies. This is because in a society where resource capture is being actively pursued, natural resource reconstruction is unlikely to occur, because it reduces the overall legitimacy of the process by politicising the issue. This is because of three fundamental factors. (A) Resource capture is basically greed driven and greed can never be satisfied. It therefore becomes self-perpetuating, *hubb al-tamalluk* (the love to possess things) as one leading shaykh put it. The notion of resource capture is also captured in the proverb *ziyadat al-khayr na'mah* (the increase of wealth is blessing). (B) Resource capture results in ecological marginalization, which mitigates against efforts at natural resource reconstruction by alienating the very people who are most affected by the resource being reconstructed. In other words, the beneficiaries are likely to be those who already have sufficient water, rather than those who have been dispossessed. (C) Resource capture, if left unattended, ultimately results in structural differentials within society. This would introduce yet another cleavage line into an already deeply divided social entity. Therefore it can be said that effective WDM needs a system of legitimate government with a prevailing normative basis, rooted in the notion of some form of equity, if it is to have a realistic chance of success.

Sixthly, while there is little evidence of any form of sustainable development the Sa'dah study does reveal potential to foster adaptive capacity. At present, what evidence exists is found in two tentative forms only. (A) There is evidence that some tribal communities are starting to learn from the mistakes of others. This is manifest in a new normative order that either discourages the further sale of land or results in tribal farmers resisting efforts by their traditional leaders to capture the resource base. (B) Religion provides another facilitating process. There is some evidence of a newly emerging interpretation of the Quran, which if continued, may result in the long-term change in perception of groundwater abstraction. A fresh religious interpretation, adjusted to modern requirements, could shift perceptions away from those justifying resource capture by a few powerful individuals, to those that are founded on notions of community and sustainability. It can be concluded that a second-order scarcity of social resources is not yet evident.

Finally, there is a need urgently to get WDM (water demand management) strategies in place if a medium-term disaster is to be avoided. Given the cultural significance of poetry, this could become an effective means of communicating WDM strategies and principles in a way that is culturally acceptable. Given the resurgence of Zaydi scholarship, new Qur'anic interpretations could provide the normative basis of these WDM strategies. There is a need for further research in order to explore these facilitating processes.

God, give us grace to accept with serenity the things that cannot be changed,
courage to change the things that should be changed,
and the wisdom to distinguish the one from the other.
(Serenity Prayer, Reinhold Niebuhr)[3]

Notes

[1] For a useful overview of this discussion see Ostrom 1990:1–26.

[2] The Zaydi resurgence is a response to the spread of Wahhabi doctrine in the area. Attacks by Saudi-trained teachers in the area have triggered responses by Sa'dah's Zaydi communities who published a little booklet called *Sa'dah limadha* (Why Sa'dah [is targeted]) (Beyrut, without date, copy held by this writer). The last page anounces a second boolet in preparation with the title *al-Wahhabiyya wa khataruha ala al-Yemen al-siyasi* (The dangers of Wahhabiyya for Yemen's political future).

[3] Composed by the German-American theologian Reinhold Niebuhr in 1943. See Bingham J. (1993) Courage to change: an introduction to the life and thought of Reinhold Niebuhr.
Internet:www.wlb-stuttgart/referate/theology/oetgeb00.html

Bibliography

Abdulfatah, K. (1981) *Mountain farmer and fellah in Asir, Southwest Arabia: the conditions of agriculture in a traditional society.* Erlanger Geographische Arbeiten, Sonderband 12, Erlangen.

Adra, N. (1985) The tribal concept in the central highlands of the Yemen Arab Republic. *Arab society: social science perspectives* (ed. by S. Eddin Ibrahim and S. Hopkins), pp. 275–285. The American University Press in Cairo, Cairo.

Alford, J. and Duguid, N. (1995) On the flatbread trail. *Aramco World,* 46, 16–25. (No 5, September-October 1995).

Allan, J.A. and Chambers, D. (1997) *Agricultural production and the trade trends: based on FAO data.* SOAS Water Issues Group, London.

Allan, J.A. and Karshenas, M. (1996) Managing environmental capital: the case of water in Israel, Jordan, the West Bank and Gaza, 1947 to 1995. *Water, peace and the Middle East: negotiating resources in the Jordan Basin* (ed. by J.A. Allan and J.H. Court), pp. 117–130. I.B. Tauris Publishers, London.

Allan, J.A. (1985) Irrigated agriculture in the Middle East: the future. *Agricultural development in the Middle East* (ed. by P. Beaumont and K.S. McLachlan), pp. 51–62. John Wiley, Chichester.

Allan, J.A. (1992) Fortunately there are substitutes for water: otherwise our hyropolitical futures would be impossible. *Water resource management,* ODA, London, pp. 13–26.

Allan, J.A. (1992) Substitutes for water are being found in the Middle East and North Africa. *Geojournal,* 28, pp. 375–385.

Allan, J.A. (1992) The importance of a realistic evaluation for non-renewable water resources in the arid and semi-arid regions. *Techniques for environmentally sound water resource development* (ed. by R.H.R. Wooldridge), pp. 63–76. Wallingford.

Allan, J.A. (1992) The Nile: the need for an integrated water management policy. *The Middle East and Europe: an integrated community approach* (ed. by G. Nonneman). Federal Trust for Education and Research, European Community DGI, London.

Allan, J.A. (1993) Economic and political adjustments to water scarcity in the Middle East. *Fifth Euro-Arab Dialogue* (ed. by C.A.O. van Niewenhuijze), pp. 43–55. Rabbani Foundation, The Hague.

Allan, J.A. (1994a) Economic and political adjustments to scarce water in the Middle East. *Water and peace in the Middle East* (ed. by J. Isaac and H. Shuval), pp. 375–388. Elsevier, Amsterdam.

Allan, J.A. (1994*b*) Food production in the Middle East. *The culinary arts of the Middle East* (ed. by S. Zubaida and R. Tapper), pp. 19–32. British Academic Press / IB Tauris, London.

Allan, J.A. (1994*c*) Mechanisms for reducing tensions over water: substituting for water in the Middle East. *MEED,* 38 (No 4, January 1994), pp. 12–14.

Allan, J.A. (1994) Overall perspectives on countries and regions. *Water in the Arab World: perspectives and progress* (ed. by P. Rogers and P. Lydon), pp. 65–100. Harvard University Press, Harvard.

Allan, J.A. (1994) Water in the Arab Middle East: availability and management. *Water as an element of cooperation and development in the Middle East* (ed. by A. Ihsan), pp. 155–200. Ayna Publications in cooperation with the Friederich Naumann Foundation in Turkey, Ankara.

Allan, J.A. (1995*a*) A transition in the political economy of water and the environment in Israel-Palestine. *Joined management of shared aquifers.* Truman Center for Peace and Palestine Consultants Group, Jerusalem.

Allan, J.A. (1995*b*) Water deficits and management options in arid regions with special reference to the Middle East and North Africa. *Water resources management in arid countries*, pp.1–8. Ministry of Water Resources, Muscat, Oman.

Allan, J.A. (1995*c*) Water in the Middle East and in Israel-Palestine: some local and global resource issues. *Joined management of shared aquifers.* Truman Center for Peace and Palestine Consultants Group, Jerusalem.

Allan, J.A. (1998) Productive efficiency and allocative efficiency: why better water management may not solve the problem. *Agricultural Water Management,* 1425, pp. 1–5. Elsevier Sciences.

Allan, J.A. (1999*a*) Contending environmental knowledge on water in the Middle East: global, regional and national contexts. Contribution to the SOAS Geography seminar on *The construction of environmental knowledge*; January-March 1999. SOAS, University of London.

Allan, J.A. (1999*b*) Global systems ameliorate local droughts: water, food and trade. *Exploratory workshop on drought and mitigation in Europe.* Space Application Institute Joint Research Centre, Ispra (Varese), Italy.

Allan, J.A. (2001) *The Middle East Water question: hydropolitics and the global economy.* IB Tauris Press, London.

Amri, H. A., al-, (1993) Yemen in the 18th and 19th centuries: the reign of Al Qasim b. Muhammad dynasty. *Studies in oriental culture and history: Festschrift for Walter Dostal* (ed. by A. Gingrich), pp. 185–198. Peter Lang, Frankfurt.

Amri, H. A., al-, (ed.) (1982) *Muhammad b. Ali al-Shawkani, al-qawl al-mufid fi adillat al-ijtihad wa-al-taqlid.* Cairo, 1340, divan al-shawkani (Aslak al-jawhar), Damascus 1402/1982.

AREA (1998) *Crop budgets – citrus, alfalfa, qat.* AREA Staff Dhamar, Yemen.

Atkinson, A. (1991) *Principles of political ecology.* Belhaven Press, London.

Badeeb, S.M. (1986) *The Saudi-Egyptian conflict over North Yemen: 1962–1970*. Westview Press, Boulder, Colorado.

Barnes, C. (1993) Water, risk and environmental management: agriculture and irrigation in South Yemen (PDRY). *New Arabian Studies*, 1, pp. 124–136.

Barrow, C. (1999) *Alternative irrigation: the promise of runoff agriculture*. Earthscan Publications Ltd., London.

Beaumont, P., Blake, G. and Wagstaff J. (1988) *The Middle East: a geographical study*, 2nd edn. David Fulton Publishers, London.

Beck, R. (ed.) (1990) *Environmental profile Al-Bayda Governorate, Yemen Arab Republic*. DHV Consultants.

Becker, H., Höhfeld, V. and Kopp, H. (eds.) (1979) *Kaffee aus Arabien: der Bedeutungswandel eines Weltwirtschaftsgutes und seine siedlungsgeographische Konsequenz an der Trockengrenze der Ökumene*. Franz Steiner Verlag, Wiesbaden.

Betzler, E. (1984) Entwicklung und Umbruch in der stadtnahen Landwirtschaft des Umlandes von San'a'. *Entwicklungsprozesse in der Arabischen Republik Jemen* (ed. by H. Kopp and G. Schweizer), pp. 55–74. Dr Ludwig Reichert Verlag, Wiesbaden.

Bingham J. (1993) *Courage to change: an introduction to the life and thought of Reinhold Niebuhr*. Internet: www..wlb-stuttg.../referate/theology/oetgeb00.html

Brunner, U. and Haefner, H. (1986) The successful floodwater farming system of the Sabeans, Yemen Arab Republic. *Applied Geography*, 6, pp. 77–86.

Brunner, U. (1985) Irrigation and land use in the Ma'rib region. *Economy, society and culture in contemporary Yemen* (ed. by B.R. Pridham), pp. 51–63. Croom Helm, London.

Brunner, U. (1999) *Jemen: vom Weihrauch zum Erdöl*. Böhlau, Wien.

Bryant, R.L. and Bailey, S. (1997) *Third World political ecology*. Routledge, London.

Bryant, R.L. (1991) Putting politics first: the political ecology of sustainable development. *Global Ecology and Biogeography Letters*, 1, pp. 164–166.

Bryant, R.L. (1997) Beyond the impasse: the power of political ecology in Third World environmental research. *Area*, 29, pp. 5–19.

Bryant, R.L. (1998) *Power, knowledge and political ecology in the Third World: a review*. Paper presented at SOAS Dept. of Geography, Departmental Seminar Series.

Burrowes, R.D. (1987) *The Yemen Arab Republic: the politics of development, 1962–86*. Westview Press, Boulder, Colorado.

Burrowes, R.D. (1995) *Historical dictionary of Yemen*. Asian Historical Dictionaries No. 17. Scarecrow Press, Lanham, Maryland.

Caponera, D. (1973) *Water laws in Muslim countries*. Irrigation and drainage paper 21/1, FAO, Rome.

Carapico, S. (1985) Yemeni agriculture in transition. *Agricultural development in the Middle East* (ed. by P. Beaumont and K.S. McLachlan), pp. 241–254. John Wiley and Sons Ltd, New York.

Carapico, S. (1998) *Civil society in Yemen: the political economy of activism in modern Arabia.* Cambridge Middle East Studies, 9, Cambridge University Press, Cambridge.

Carnap, M. and Beier (1996) *Comments on the discussion papers.* Yemen: water strategy, Dec.27 1995.

Caton, S.C. (1986) Salam tahiyah: greetings from the highlands of Yemen. *American Ethnologist,* 13, pp. 290–308.

Caton, S.C. (1990a) *Peaks of Yemen I summon: poetry as cultural practice in a North Yemeni tribe.* University of California Press, Berkeley.

Caton, S.C. (1990b) Anthropological theories of tribe and state formation in the Middle East: ideology and the semiotics of power. *Tribes and state formation in the Middle East* (ed. by P.S. Khouri and J. Kostiner), pp. 74–108. University of California Press, Berkeley.

Central Planning Organization, *The Second Five-Year Plan, 1982–1986, Yemen Arab Republic.*

Chapman, J. (1990) The carving of the rhino janbiyya hilt in North Yemen. *Arabian Studies,* 8, pp. 11–21.

Chaudhry, K.A. (1997) *The price of wealth: economies and institutions in the Middle East.* Cornell University Press, Ithaca and London.

Chaudry, M.A. (1991) *Estimating water requirements for efficient water sector planning: projected estimates for the agricultural sector.* (Seminar on economic and social issues in water resources development and management in Yemen) T.S. of the H.W.C. San'a.

Chaudry, M.A., Turkawi, A.G. and Tejada, J.A. (1990) *Regional agricultural water requirements in the northern part of the Yemen Republic.* T.S. of the H.W.C. (YEM-88–001), San'a.

Cockburn, A. (2000) Yemen. *National Geographic,* (April 2000), pp. 34–53.

Danikh, M. and Van der Gun, J.A.M. (1985) *Water resources of the Sadah area.* Annex 3: Hydrological network. Report WRAY-3.3. YOMINCO/TNO, San'a/ Delft.

Davis, J.P. and Garvey, G. (1993) *Developing and managing community water supplies.* Oxfam Development Guidelines No.8, Oxfam Publications, Oxford.

Detalle, R. (2000) Saudi Arabia and Yemen: beyond the boundaries. *Tensions in Saudi Arabia: the Saudi-Yemeni fault line* (ed. by Renaud Detalle), pp. 150–69. Aktuelle Materialien zur Internationalen Politik, Stiftung Wissenschaft und Politik 7 SWP, Conflict Prevention Network (SWP-CPN). Nomos Verlagsgesellschaft, Baden-Baden.

Detalle, R. (2000) The Yemeni-Saudi conflict: bilateral transactions and interactions. *Tensions in Saudi Arabia: the Saudi-Yemeni fault line* (ed. by Renaud Detalle), pp. 52–79. Aktuelle Materialien zur Internationalen Politik,

Stiftung Wissenschaft und Politik 7 SWP, Conflict Prevention Network (SWP-CPN). Nomos Verlagsgesellschaft, Baden-Baden.

DHV (1992) *Groundwater resources and use in the Sa'dah plain.* Draft Report NORADEP. Prepared for SSHARDA under UNDP project YEM/87/015. DHV (The Neth.) in association with Team Consulting and Darwish Cons. Engineers.

DHV (1993*a*) *Agro-economic and sociological sondeo study* (volume 3: target area field reports, Sa'dah province, Sa'dah plain). NORADEP. Prepared for SSHARDA under UNDP project YEM/87/015. DHV (The Neth.) in association with Team Consulting and Darwish Cons. Engineers.

DHV (1993*b*) *Water management plan: Sa'dah plain target area.* Final Report NORADEP. Prepared for SSHARDA under UNDP project YEM/87/015. DHV (The Neth.) in association with Team Consulting and Darwish Cons. Engineers.

Dobert, M. (1984) Development of aid programs in Yemen. *American Affairs,* 8 (Spring 1994), pp. 108–116.

Dorski, S. (1985) *Women of Amran: a Middle Eastern ethnographic study.* University of Utah Press, Salt Lake City.

Dresch, P. (1981) The several peaces of Yemeni tribes. *Journal of the Anthropological Society of Oxford,* 7, pp. 73–86.

Dresch, P. (1984) Tribal relations and political history in Upper Yemen. *Contemporary Yemen: politics and historical background* (ed. by B.R. Pridham), pp. 154–174. Croom Helm, London.

Dresch, P. (1984) Tribal relations and political history in Upper Yemen. *Contemporary Yemen: politics and historical background* (ed. by B.R. Pridham), pp. 154–174. Croom Helm, London.

Dresch, P. (1989) *Tribes, government, and history in Yemen.* Clarendon Press, Oxford.

Dresch, P. (1990) Imams and tribes: the writing and acting of history in Upper Yemen. *Tribes and state formation in the Middle East* (ed. by P.S. Khouri and J. Kostiner), pp. 252–287. University of California Press, Berkeley.

Dresch, P. (1993) A daily plebiscite: nation and state in Yemen. *Revue du Monde Musulman et de la Mediterranee,* 67, pp. 67–77.

Dresch, P. (1995) The tribal factor in the Yemeni crisis. *The Yemeni war of 1994: causes and consequences* (ed. by J.S. al-Suwaidi), pp. 33–56. Saqi Books, London.

Dresch, P. and Haykel B. (1995) Stereotypes and political styles: Islamists and tribesfolk in Yemen. *IJMES,* 27, pp. 405–431.

Dubach, W. (1977) *Arab Republic of Yemen: a study of traditional forms of habitation and types of settlement.* Airphoto Interpretation Project, Zürich.

Eagle, A.B.D.R. (1994) Al-Hadi Yahyà b. al-Husayn b. al-Qasim (245–98/859–911): a biographical introduction of the background and significance of his Imamate. *New Arabian Studies,* 2.

EC (European Commission) (1998) *Towards sustainable water resources management: a strategic approach.* Guidelines for water resources development co-operations. ECSC-EEC-EAEC, Brussels, Luxembourg.

Eger, H. (1984) Rainwater harvesting in the Yemeni highlands: the effect of rainwater harvesting on soil moisture status and its implications for arable farming, a case study of the Amran region. *Entwicklungsprozesse in der Arabischen Republik Jemen* (ed. by H. Kopp and G. Schweizer), pp. 147–170. Dr. Ludwig Reichert Verlag, Wiesbaden.

Eger, H. (1986) *Runoff agriculture: a case study about the Yemeni highlands.* Dr. Ludwig Reichert Verlag, Wiesbaden.

El-Azhari, M.S. (1984) Aspects of North Yemen's relations with Saudi Arabia. *Contemporary Yemen: politics and historical background* (ed. by B.R. Pridham), pp. 195–207. Croom Helm, London.

El-Daher, S. and Geissler C. (1990) North Yemen: from farming to foreign funding. *Food Policy,* 15, pp. 531–535

Elderhorst, W. and Van der Gun, J.A.M. (1985) *Water resources of the Sadah area.* (Annex 7, groundwater availability, report WRAY-3.7).

Elewaut, E. (1985) *Water resources of the Sadah area.* (Annex 5, geophysical well-logging, report WRAY-3.5).

Emiel, J., Roberts, R. and Sauri D. (1992) Ideology, poverty, and groundwater resources: an exploration of relations. *Political Geography,* 11 (No.1), pp. 37–54.

Eryani, M.L., al-, Bamatraf, A., Saqaf, G., al-, and Hadash, S. (1995) *Water right aspects of the proposed sources of additional water supply for the city of Sana'a.* Sana'a, Republic of Yemen.

FAO (1977) *Crop water requirements.* FAO irrigation and drainage paper no. 24, FAO, Rome.

Ferguson, J. (1994) *The anti-politics machine: "development," depoliticization, and bureaucratic power in Lesotho.* University of Minnesota Press, Minneapolis.

Findlay, A. (1994) *The Arab world.* Routledge, London.

Fischer, W. (1984) The economic and sociogeograghic effects of labour migration in two villages in the Yemen Arab Republic. *Entwicklungsprozesse in der Arabischen Republik Jemen* (ed. by H. Kopp and G. Schweizer), pp. 99–118. Dr. Ludwig Reichert Verlag, Wiesbaden.

Fisher, F.M. (1995) Water and peace in the Middle East. *Middle East International,* 17 (November 1995), pp. 17–18.

Forrer, L. (1942) *Südarabien nach al-Hamdanis "Beschreibung des Morgenlandes"* Heft 27,3, Leipzig 1942, Reprint Nendels 1966.

Frese-Weghöft, G. (1986) *Ein Leben in der Unsichtbarkeit: Frauen im Jemen.* RoRoRo Aktuell 5645.

Gamal, N., Qadir, N., and Van der Gun, J.A.M. (1985) *Water resources of the Sadah area.* Technical annex 1: Well inventory results. Report WRAY-3.1, YOMINCO/TNO, San'a/Delft.

Gause, F.G. (1990) *Saudi-Yemeni relations: domestic structures and foreign influence.* Columbia University Press, New York.

Gebhardt, H. and Schweizer, G. (1992) Märkte in einem Entwicklungsland im wirtschaftlichen Umbruch: das Beispiel der Republik Jemen. *Die Erde,* 123, pp. 95–112.

Gerholm, T. (1977) *Market, mosque and mafraj: social inequality in a Yemeni town.* Stockholm Studies in Social Anthropology 5, Stockholm.

Gingrich, A. and Heiss, J. (1986) *Beiträge zur Ethnographie der Provinz Sa'da (Nordjemen): Aspekte der traditionellen materiellen Kultur in bäuerlichen Stammesgesellschaften.* Österreichische Akademie der Wissenschaften, Wien.

Gingrich, A. (1993) Tribes and rulers in northern Yemen. *Studies in oriental culture and history: Festschrift for Walter Dostal* (ed. by A. Gingrich), pp. 253–280. Peter Lang, Frankfurt.

Grohmann, A. and Irvine, A.K. (1960) Khawlan. *Encyclopedia of Islam,* new edn. Leiden.

Grohmann, A. (1930–1934) *Südarabien als Wirtschaftsgebiet.* Verlag des Forschungsinstitutes für Osten und Orient, Brunn, Rudolf, Wien.

Haddash, S. (1996) Water legislation in the Republic of Yemen. *Symposium on water and Arab gulf development: problems and policies.* University of Exeter, Centre for Arab Gulf studies.

Haddash, S. (1998) Water diversion rights. *Yemen Times,* March 2 1998.

Haddash, S. (1998) Water ownership rights. *Yemen Times,* February 23 1998.

Handley, C. and Saqqaf G., al-, (1996) *Water use patterns in the Amran valley, Yemen.* Yemen water strategy (decentralised management study, Amran component), The World Bank.

Handley, C. (1996) *Baseline socio-economic survey in Ta'iz governorate (excluding Tihama),* UNDP/DDSMS Project YEM/93/010.

Handley, C. (2001) Water stress: some symptoms and causes: a case study from Ta'iz, Yemen. Ashgate, Aldershot.

Hardin, G. (1968) The Tragedy of the Commons. *Science,* p. 126, pp. 1243–1248.

Hartmann, R. (1995) Yemeni exodus from Saudi Arabia: the Gulf conflict and the ceasing of the worker's emigration. *Journal of South Asian and Middle East Studies,* 19 (2), pp. 38–52.

Haykel, B. (1993) Al-Shawkani and the jurisprudential unity of Yemen. *Revue du Monde Musulman et de la Mediterranee,* p. 67, pp. 53–65.

Haykel, B. (1995) A Zaydi revival? *Yemen Update,* p. 36, pp. 20–21 (Bulletin of the American Institute of Yemeni Studies).

Heiss, J. (1987) Historical and social aspects of Sa'dah, a Yemeni town. *Proceedings of the Seminar for Arabian Studies,* 17, pp. 63–79.

Held, C. (1989) *Middle Eastern Patterns.* Westview Press, San Francisco.

Hoekstra, A. (1998) *Perspectives on water: an integrated model-based exploration of the future.* International Books, Utrecht.

Höhfeld, V. (1978) Die Entwicklung der administrativen Gliederung und die Verwaltungszentren in der Arabischen Republik Jemen (Nordjemen). *Orient*, 2, pp. 22–63.

Hossain, M.M. and Nouman, S. (1991) *Agro-economic survey in selected state farms Qa' al-Boun, Hatarish and Jihana.* UNDP/FAO/YEM/87/001.

IDRC (1998) (International Development Research Centre) *Water resources management in the Islamic World.* Conference report, December 1–3, 1998, Amman, Jordan.

IIMI (1996) *Mission statement of the International Irrigation Management Institute.* IIMI, Colombo.

Joffe, E., Hachemi, M. and Watkins E. (1997) *Yemen today: crisis and solutions.* Proceedings of a two-day conference held at the School of Oriental and African Studies, University of London November 25th and 26th, 1995. Caravel Press, London.

Johnston, R.J. (1994) *Human Geography.* 3rd Edn. Blackwell Publishers, Oxford.

Jungfer, E.V. (1984) Das Wasserproblem im Becken von San'a'. Anthrogene und physische Ursachen einer zunehmenden Austrocknung. *Entwicklungsprozesse in der Arabischen Republik Jemen* (ed. by H. Kopp and G. Schweizer), pp. 171–94. Dr. Ludwig Reichert Verlag, Wiesbaden.

Kasir, A., al-, (1985) The impact of emigration on the social structure in the Yemen Arab Republic. *Economy, society and culture in contemporary Yemen* (ed. by B.R. Pridham), pp. 122–131. Croom Helm, London.

Kennedy, S. (1998) *Yemen: a pictorial tour.* Motivate Publishing, London.

Kennedy, John G. (1987) *The flower of paradise: the institutionalized use of the drug Qat in North Yemen.* D. Reidel Publishing Company, Dordrecht, The Netherlands.

Khouri, P.S. and Kostiner, J. (1990) Introduction: tribes and the complexities of state formation in the Middle East. *Tribes and state formation in the Middle East* (ed. by P.S. Khouri and J. Kostiner), pp. 1–22. University of California Press, Berkeley.

Kohler, S. (1999) *Institutionen in der Bewässerungs-Landwirtschaft im Jemen: Die Ursachen der Wasserübernutzung.* Jemen-Studien, Band 13. Dr Ludwig Reichert Verlag, Wiesbaden.

Kohler, S. (1999) Hintergründe zur Wasserkrise im Jemen. *Jemen-Report*, 30/2, pp. 4–10.

Kohler, S. (1999) Die Wasserübernutzung im Jemen: Institutionelle Ursachen und Lösongsmöglichkeiten. *Zeitschrift für Bewässerungswirtschaft*, 34/2, pp. 109–123.

Kopp, H. and Aves, A.M. (eds.) (1988) *Jemen (Nord) im Aufbruch. Tendenzen der wirtschaftlichen, kulturellen und politischen Entwicklung.* Tübingen/Bonn.

Kopp, H. (1977) *Al Qasim. Wirtschafts – und sozialgeographische Strukturen und Entwicklungsprozesse in einem Dorf des jemenitischen Hochlandes.* (Beihefte zum Tübinger Atlas des Vorderen Orients Nr. 31). Dr. Ludwig Reichert, Wiesbaden

Kopp, H. (1981) *Agrargeographie der Arabischen Republic Jemen. Landnutzung und agrarsoziale Verhältnisse in einem Islamisch-orientierten Entwicklungsland mit alter bäuerlicher Kultur.* Erlanger Geographische Arbeiten, Sonderband 11, Erlangen.

Kopp, H. (1985) Land usage and its implications for Yemeni agriculture. *Economy, Society and Culture in Contemporary Yemen* (ed. by B.R. Pridham), pp. 41–50. Croom Helm, London.

Kopp, H. (2000) Water and mineral resources in Yemen and Saudi Arabia. *Tensions in Saudi Arabia: the Saudi-Yemeni fault line* (ed. by Renaud Detalle), pp. 80–95. Aktuelle Materialien zur Internationalen Politik, Stiftung Wissenschaft und Politik 7 SWP, Conflict Prevention Network (SWP-CPN). Nomos Verlagsgesellschaft, Baden-Baden.

Koszinowski, T. (1993) Abdallah Ibn Hussain al-Ahmar. *Orient*, p. 34, pp. 335–41.

Koszinowski, T., El-Menshaui, M. and Meyer A. (1980) *Zur politischen und wirtschaftlichen Situation des Yemen.* Einführung und Dokumentation. Deutsches Orient Institut, Hamburg.

Kruse, H. (1981) Verwaltungsgeschichte und Sozialgeschichte – dargestellt am Beispiel der Arabischen Republick Jemen. *Perpektiven der Entwicklungspolitik* (ed. by K. Ringer, E.A. Renesse und C. Uhlig), pp. 185–213. Horst Erdmann Verlag, Tübingen.

Leedy, P.D. (1997) *Practical research: planning and design.* Sixth edition. Simon and Schuster, New Jersey.

Lichtenthäler, G. and Turton, A.R. (1999) *Water demand management, natural resource reconstruction and traditional value systems: a case study from Yemen. MEWREW* Occasional Paper No.14. SOAS Water Issues Group, University of London. Available on website: www2.soas.ac.uk/Geography/WaterIssues/OccasionalPapers

Lichtenthäler, G. (1996) Tribes and trends: changing perceptions of the value of water in Yemen. *Perceptions of the value of water and water environments,* Proceedings of the European seminar on water geography 6–10 September 1996, London, SOAS, University of London.

Lichtenthäler, G. (1999) *Water management and community participation in the Sa'dah basin of Yemen: politics, perceptions and perspectives.* The World Bank.

Lichtenthäler, G. (2001) Die Politisierung der natürlichen Ressourcen im Sa'dah-Becken. *Jemen-Report,* 32/2, pp. 25–28.

Lichtenthäler, G. (2001) Politik, Power und Prestige: Gesellschaftliche Werte des Wassers im Jemen. *inamo,* 27/2001.

Lichtenthäler, G. (2001) Power, politics and patronage: adaptation of water rights among Yemen's northern highland tribes. *Etudes Rurales, pp.* 155–156 (juillet-december 2000), pp. 143–166.

Lichtenthäler, G. (2001) *Sa'dah limadha* - Warum Sa'dah? Wassermanagement in der Probe. *Jemen-Report,* 32/1, pp. 12–18.

Lloyd-Jones, D. Martyn (1981) *Spiritual depression: its causes and its cure.* Eerdmans, Michigan.

Lundqvist, J. and Gleick, P. (1997) *Comprehensive assessment of the freshwater resources of the world: sustaining our waters into the 21st century.* Stockholm Environmental Institute, Stockholm.

Mackintosh-Smith, T. (1997) *Yemen-travels in dictionary land.* John Murray, London.

Madelung, W. (1984) Land ownership and land tax in northern Yemen and Najran: 3rd-4th/9th-10th century. *Land tenure and social transformation in the Middle East* (ed. by T. Khalidi), pp. 189–207. American University of Beirut, Beirut.

Maktari, A. (1971) *Water rights and irrigation practices in Lahj.* Cambridge University Press, Cambridge.

Meinhold, K.D. and Trurnit, P. (1981) *Prospektion und Exploration von bekannten Erzvorkommen im Raum Sadah.* (Endbericht, Teil I), Bundesanstalt für Geowissenschaften und Rohstoffe, Hannover.

Merrett, S. (1997) *Introduction to the economics of water resources: an international perspective.* UCL Press, London.

Messick, B.M. (1978) *Transactions in Ibb: economy and society in a Yemeni highland town.* Ph.D. thesis, University Microfilms International, Ann Arbor, Michigan, USA.

Meyer, G. (1985a) Sozioökonomische Handlungsstrategien und sozialgruppen-spezifische Kooperationsformen im informellen Sector von Sanaa/Nordjemen. *Zeitschrift für Wirtschaftsgeographie,* 29, pp. 107–116.

Meyer, G. (1985b) Labour emigration and internal migration in the Labour emigration and internal migration in the Yemen Arab Republic – the urban sector. *Economy, society and culture in contemporary Yemen* (ed. by B.R. Pridham), pp. 147–64. Croom Helm, London.

Meyer, G. (1986) *Arbeitsemigration, Binnenwanderung und Wirtschaftsentwicklung in der Arabischen Republik Jemen.* Jemen-Studien, Band 2. Dr. Ludwig Reichert Verlag, Wiesbaden.

Migdal, J. (1988) *Strong societies and weak states: state-society relations and state capabilities in the Third World.* Princeton University Press, Princeton.

Milroy, A. (1999) *Renewing Yemen's traditional capacity for local community development through ta'awun, tribe and modern agricultural associations.* Arid Lands Initiative, www.al-bab.com/yemen/env/arid.htm

Mitchell, D. (1995) *The wheat subsidy in Yemen.* Commodity and Analysis Unit, International Economics Department, The World Bank.

Moench, M. (1997) *Local water management: options and opportunities in Yemen.* Summary report to the World Bank of the Decentralized Management Study. The World Bank, Republic of Yemen.

Mohieldeen, Y. (1999) *Responses to water scarcity: social adaptive capacity and the role of environmental information. A case study from Tai'z, Yemen.* MA dissertation, SOAS, University of London. www2.soas.ac.uk/Geography/WaterIssues/OccasionalPapers

Morris, J. (1996) Water policy: economic theory and political reality. *Water policy: allocation and management in practice* (ed. by P. Howsam). E and FN Spon, London.

Morris, T. (1986) *Adapting to wealth: social changes in a Yemeni highland community.* Ph.D. thesis, SOAS, University of London.

Morris, T. (1991) *The despairing developer: diary of an aid worker in the Middle East.* I.B.Tauris, London.

Mortimore, M. (1989) *Adapting to drought: farmers, famines and desertification in West Africa.* Cambridge University Press, Cambridge.

Morton, J. (1994) *The poverty of nations: the aid dilemma at the heart of Africa.* British Academic Press, London.

Mostert, E. (1998) A framework for conflict resolution. *Water International* 23 (No. 4), pp. 206–215.

Nichols, P. (1991) *Social survey methods: a verification for development workers.* Development guidelines 6, Oxfam, Oxford.

Niewöhner-Eberhard, E. (1976) Täglicher Suq und Wochenmarkt in Sa'da, Jemen. *Erdkunde*, 30, pp. 24–27.

Niewöhner-Eberhard, E. (1977) Das jemenitisch-arabische Innenhofhaus in Sa'da, Jemen. *Der Islam*, 54, pp. 177–204.

Niewöhner-Eberhard, E. (1985) *Sa'da. Bauten und Bewohner in einer traditionellen Islamischen Stadt.* Dr. Ludwig Reichert Verlag, Wiesbaden.

Niewöhner-Eberhard, E. (1987) *Veränderungen im Marktangebot. Der Suq von Sa'da in den Jahren 1973 und 1983.* Baessler Archiv NF 35:357–81.

O'Ballance, E. (1971) *The war in the Yemen.* Faber and Faber, London.

Ohlsson, L. (1998) *Water and social resource scarcity – an issue paper commissioned by FAO/AGLW.* Presented as a discussion paper for the 2nd FAO e-mail conference on managing water scarcity. WATSCAR 2.

Ohlsson, L. (1999) *Environment, scarcity and conflict: a study of Malthusian concerns.* Department of Peace and Development Research, Göteborg University.

Ostrom, E. (1990) *Governing the commons: the evolution of institutions for collective action.* Cambridge University Press, Cambridge.

Peterson, J.E. (1985) The islands of Arabia: their recent history and strategic importance. *Arabian Studies*, 7, 23–35.

Poate, C.D. and Daplyn, P.F. (1993) *Data for agrarian development.* Cambridge University Press, Cambridge.

Postel, S. (1999) *Pillars of sand: can the irrigation miracle last?* Norton and Company, London.

Prebble, J. (1969) *The Highland Clearances.* Penguin, London.

Puin, G-R. (1984) The Yemeni *hijra* concept of tribal protection. *Land tenure and social transformation in the Middle East.* (ed. by T. Khalidi), pp. 483–93. American University of Beirut, Beirut.

Quist, D. (1990) *Im Norden des Jemen. Vom Roten Meer zur arabischen Wüste.* Harenberg Edition, Dortmund.

Raban, J. (1979) *Arabia through the looking glass.* Collins, London.

Redclift, M. and Benton T. (eds.) (1994) *Social theory and the global environment.* Routledge, London.

Redclift, M. (1984) *Development and the environmental crisis: red or green alternatives.* Methuen, London.

Redclift, M. (1987) *Sustainable development: exploring the contradictions.* Methuen, London.

Redclift, M. (1992) Sustainable development and popular participation: a framework for analysis. *Grassroots environmental action: people's participation in sustainable development* (ed. by D. Ghai and J.M. Vivian), pp. 23–49. Routledge, London.

Reisner, M. (1993) *Cadillac Desert: The American West and its disappearing water.* Revised Edition, Penguin, New York.

Republic of Yemen *Statistical Year-Book 1995.* Ministry of Planning and Development, Central Statistical Organization.

Republic of Yemen (1996) *Population and Housing Census Dec-1994.* Ministry of Planning and Development.

Richards, A. and Waterbury, J. (1990) *A political economy of the Middle East: state, class, and economic development.* Westview Press, Boulder, Colorado.

Roberts, J. (1995) *Visions and mirages: Middle East in New Era.* Mainstream Publishing Company, Edinburgh.

Rosenau, J.N. (1993) Thinking theory thoroughly. *International relations theory* (ed. by P. Notti and M. Hoppi) 2nd edn. Macmillan.

Saidi, M.A., al-, (ed.) (1993) *The cooperative movement of Yemen and issues of regional development.* Klaus Schwartz Verlag, Berlin.

Sakkaf, Z.A. al-, Juned A. and Tracey White J. (1998) Review of marketing policies and analysis of the strategies. *Yemen: agricultural policy review.* Working paper number 2. The World Bank, Ministry of Agriculture and Irrigation, San'a.

Sakkaf, R.A. al-, Zhou, Y. and Hall, M. (1999) A strategy for controlling groundwater depletion in the Sa'dah. *Water Resources Development,* 15, pp. 349–365.

Sayer, A. (1984) *Methods in social science: a realist approach.* Hutchinson and Co Ltd, London.

Schofield, R. (2000) The international boundary between Yemen and Saudi Arabia. *Tensions in Arabia: the Saudi-Yemeni fault line* (ed. by Renaud Detalle), pp. 15–51. Aktuelle Materialien zur Internationalen Politik, Stiftung Wissenschaft und Politik 7 SWP, Conflict Prevention Network (SWP-CPN). Nomos Verlagsgesellschaft, Baden-Baden.

Schopen, A. (1978) *Das Qat. Geschichte und Gebrauch des Genussmittels Catha Edulis Forsk in der Arabischen Republik Jemen.* Franz Steiner Verlag, Wiesbaden.

Schweizer, G. (1984) Wochenmärkte in der Arabischen Republic Jemen. Das traditionelle Versorgungssystem unter dem Einfluß von Entwicklungsprozessen. *Entwicklungsprozesse in der Arabischen Republik Jemen* (ed. by H. Kopp and G. Schweizer), pp. 9–26. Dr. Ludwig Reichert Verlag, Wiesbaden.

Schweizer, G. (1985) Social and economic change in the rural distribution system: weekly markets in the Yemen Arab Republic. *Economy, society and culture in contemporary Yemen* (ed. by B.R. Pridham), pp. 107–121. Croom Helm, London.

Serjeant, R.B. (1995) *Farmers and fishermen in Arabia: studies in customary law and practice* (ed. by G.R. Smith). Variorum, Collected Studies series.

Serjeant, R.B. (1995) Some irrigation systems in the Hadramawt. *Farmers and fishermen in Arabia: studies in customary law and practice* (ed. by G.R. Smith). Variorum, Collected Studies series.

Steffen, H. (1979) *Population geography of the Yemen Arab Republic: the major findings of the population and housing census of February 1979.* Dr. Ludwig Reichert, Wiesbaden.

Stookey, R.W. (1974) Social structure and politics in the Yemen Arab Republic. *Middle East Journal,* 28, pp. 248–260 and pp. 409–418.

Stookey, R.W. (1978) *Yemen: the politics of the Yemen Arab Republic.* Westview Press, Boulder, Colorado.

Stookey, R.W. (1984) Yemen: revolution versus tradition. *The Arabian Peninsula: zone of ferment* (ed. by R.W. Stookey), pp. 79–108. Hoover Institution Press, Stanford.

Stott, J. (1984) *Issues facing Christians today.* Marshalls, London.

Suliman, M. (1993) Krieg um schwindende Lebensgrundlagen: Der Bürgerkrieg im Sudan als ökologischer Konflikt. *Der Überblick,* 3/93, pp. 59–62.

Swagman, C.F. (1988) *Development and change in highland Yemen.* University of Utah Press, Salt Lake City, Utah.

Swanson, J.C. (1979) Some consequences of emigration for rural economic development in the Yemen Arab Republic. *The Middle East Journal,* 33, 34–43.

The Government of the Republic of Yemen, High Water Council (1992) *Water resources management in the context of national economic development.* Volume I. UNDP/DESD Project YEM/88/001, San'a, Republic of Yemen.

The Government of the Republic of Yemen, High Water Council (1992) *National water legislation and institutional issues.* Volume II. UNDP/DESD Project YEM/88/001, San'a, Republic of Yemen.

The Government of the Republic of Yemen, High Water Council (1992) *Surface water resources.* Final report, volume III. UNDP/DESD Project YEM/88/001, San'a, Republic of Yemen.

The Government of the Republic of Yemen, High Water Council (1992) *Groundwater resources.* Final report, volume IV. UNDP/DESD Project YEM/88/001, San'a, Republic of Yemen.

The Government of the Republic of Yemen, High Water Council (1992) *Regional water requirements for different water-consuming sectors.* Final report, volume V. UNDP/DESD Project YEM/88/001, San'a, Republic of Yemen.

The Government of the Republic of Yemen, High Water Council (1992) *Water supply, wastewater and sanitation.* Final report, volume VI. UNDP/DESD Project YEM/88/001, San'a, Republic of Yemen.

Thenayian, M.A.R., al-, (1997) A preliminary evaluation of Al-Rada'i's Urjuzat al-Hajj as a primary geographical source for surveying the Yemeni highland pilgrim route. *New Arabian Studies* 4, pp. 243–260.

Thompson, M. (1995) Policy-making in the face of uncertainty: the Himalayas as unknowns. *Water and the quest for sustainable development in the Ganges Valley* (ed. by G.P. Chapman and M. Thompson), pp. 25–38. Mansell Publishing, London.

Turton, A.R. and Ohlsson, L. (1999) *Water scarcity and social adaptive capacity: towards an understanding of the social dynamics of managing water scarcity in developing countries.* Urban stability through integrated water-related management. The 9th Stockholm Water Symposium, 9–12 August 1999, Stockholm. www2.soas.ac.uk/Geography/WaterIssues/OccasionalPapers

Turton, A.R. (1999) *Water scarcity and social adaptive capacity: towards an understanding of the social dynamics of water demand management in developing countries.* MEWREW Occasional Paper No. 9. Water Issues Study Group, School of Oriental and African Studies (SOAS), University of London. www2.soas.ac.uk/Geography/WaterIssues/OccasionalPapers

Tutwiler, R.N. and Carapico, S.H. (1981) *Yemeni agriculture and economic change: case studies of two highland regions.* American Institute for Yemeni Studies, San'a'/Milwaukee.

UNDP (1995) *Yemen water strategy: discussion papers.* UNDP/The Netherlands.

Unwin, T. (ed.) (1994) *Atlas of world development.* John Wiley and Sons, Chichester.

Van der Gun, J.A.M. and Ahmed, A.A. (eds.) (1995) *The water resources of Yemen: a summary and digest of available information.* Report WRAY-35, Ministry of Oil and Mineral Resources, General department of Hydrology/TNO Institute of Applied Geoscience, San'a/Delft.

Van der Gun, J.A.M. (1983) *Interim report on the water resources of the Sadah area.* Report WRAY-2, YOMINCO/TNO, San'a/Delft.

Van der Gun, J.A.M. (1985) *Water resources of the Sadah area: main report.* Report WRAY-3, YOMINCO/TNO, San'a'/Delft.

Van der Gun, J.A.M. (1987) *The Sadah study in retrospective: some remarks on the analysis of groundwater availability.* Internal note, WRAY project.

Varisco, D.M. (1982) *The adaptive dynamics of water allocation in al-Ahjur, Yemen Arab Republic.* Ph.D. thesis, University Microfilms International, Ann Arbor, Michigan, USA.

Varisco, D.M. (1982) The *ard* in highland Yemeni agriculture. *Tools and Tillage*, 4, pp. 158–172.

Varisco, D.M. (1983*a*) Irrigation in an Arabian valley: a system of highland terraces in the Yemen Arab Republic. *Expedition*, 25, pp. 26–34.

Varisco, D.M. (1983*b*) *Sayl* and *ghayl*: the ecology of water allocation in Yemen. *Human Ecology*, 11, pp. 365–383.

Varisco, D.M. (1986) On the meaning of chewing: the significance of qat (Catha Edulis) in the Yemen Arab Republic. *International Journal of Middle East Studies*, 18, pp. 1–13.

Varisco, D.M. (1993) A Rasulid agricultural almanac for 808/1405–6. *New Arabian Sudies*, 1, pp. 108–123.

Varisco, D.M. (1994) *Medieval agriculture and Islamic science: the almanac of a Yemeni sultan.* University of Washington Press, Washington, DC.

Varisco, D.M. (1996) Water sources and traditional irrigation in Yemen. *New Arabian Studies*, 3, pp. 238–257.

Vincent, L. (1990) *The politics of water scarcity: irrigation and water supply in the mountains of the Yemen Republic.* Irrigation Management Network Paper 90/3e, Overseas Development Institute, London.

Vincent, L. (1991) Debating the water decade: a view from the Yemen Republic. *Development Policy Review*, 9, pp. 197–216.

Wald, P. (1980) *Der Jemen: Antikes und islamisches Südarabien -Geschichte, Kultur und Kunst zwischen Rotem Meer und Arabischer Wüste.* DuMont, Köln.

Ward, C. (ed.) (1998) *Yemen: local water management in rural areas – a case study.* The World Bank, Republic of Yemen.

Ward C., Thawr A., Qasem F. and No'man A. (1998) Qat. Yemen: agricultural strategy. Working paper number 8. The World Bank, Ministry of Agriculture and Irrigation, San'a.

Wehr, H. (1976) *A dictionary of modern written Arabic* (ed. by J. Milton Cowan). 3rd Edn. Otto Harrassowitz, Wiesbaden.

Weir, S. (1985*a*) *Qat in Yemen: consumption and social change.* British Museum Publications, London.

Weir, S. (1985*b*) Economic aspects of the qat industry in north-west Yemen. *Economy, society and culture in contemporary Yemen* (ed. by B.R. Pridham), pp. 64–82. Croom Helm, London.

Weir, S. (1986) Tribe, *hijrah* and *madinah* in north-west Yemen. *Middle Eastern cities in perspective: points de vue sur villes du Magreb et du Machrek* (ed. by K. Brown, M. Jole, P. Sluglett and S. Zubaida), pp. 225–239. (Franco-British symposium on the comparative analysis of Arab and Muslim cities) Ithaca Press, London.

Weir, S. (1987) Labour migration and key aspects of its economic and social impact on a Yemeni highland community. *The Middle Eastern village: changing economic and social relations* (ed. by R. Lawless), pp 273–96. Croom Helm, London.

Weir, S. (1997) A clash of fundamentalisms: Wahhabism in Yemen. *Middle East Report* (July – September 1997).

Weir, S. (1998) Are Yemeni tribes disorderly, violent and against states. *Yemen: the challenge of social and economic development in the era of democracy.* Conference proceedings, University of Exeter: Centre for Arab Gulf Studies, 1st-4th April, 1998.

Weiter, M. (1987) Development and development aid in Yemen. *Yemen: 3000 years of art and civilisation in Arabia Felix* (ed. by W. Daum), pp. 419–423. Pinguin-Verlag, Innsbruck.

Wenner, M. (1991 The Yemen Arab Republic: development and change in an ancient land. Westview Press, Boulder, Colorado.

Westerlund, H. (1996) To meter or not to meter, that is the question. *Water policy: allocation and management in practice* (ed. by P. Howsam). E and FN Spon, London.

Wilkinson, J.C. (1977) *Water and tribal settlement in south-east Arabia: a study of the aflaj of Oman.* Clarendon Press, Oxford.

Wilkinson, J.C. (1978) Islamic water law with special reference to oasis settlement. *Journal of Arid Environments*, 1, pp. 87–96.

Winpenny, J.T. (1994) *Managing water as an economic resource.* Routledge, London.

Winpenny, J.T. (1997) Demand management for efficient and equitable use. *Water: economics, management and demand* (ed. by K.M. Franks and L. Smith). E and FN Spon, London.

World Bank (1979) *Yemen Arab Republic: development of a traditional economy.* The World Bank, Washington, DC.

World Bank (1986) *Yemen Arab Republic agricultural strategy paper.* (Report no. 5574), Yemen Arab Republic/The World Bank.

World Bank (1993) Agricultural sector study strategy for sustainable agricultural development. The World Bank.

World Bank (1994) *A strategy for managing water in the Middle East and North Africa.* The World Bank, Washington DC.

World Bank (1997a) *Local water management: options and opportunities in Yemen.* Summary report to the World Bank of the decentralized management study. The World Bank, San'a.

World Bank (1997b) *Yemen: towards a water strategy. An agenda for action.* The World Bank.

World Bank (1998) *Republic of Yemen: agricultural strategy note.* Draft copy. The World Bank.

Zagni, A.F.E. (1996) *Study of cropping systems, water requirements, and irrigation practices in Amran valley (draft).* Yemen water strategy (decentralized management study), The World Bank.

Maps

Survey Authority San'a (1983) *The Yemen Arab Republic* 1:5,000 (Maps, 4 sheets of Sa'dah town and immediate vicinity.

Survey Authority San'a (1985) *The Yemen Arab Republic* 1:500,000 (Maps, two sheets) Produced for the YAR by the Director of Military Survey, Ministry of Defence, UK.

Survey Authority San'a (1989) Maps 1:50,000 (4 sheets of Sa'dah area: Sa'dah; Wadi Nashur; Majz; Saqayn) Arabic version.

Satellite Images

Landsat TM, Path: 166, Row: 48, Date: 20 August 1998.

Index

Printed and bound by CPI Group (UK) Ltd, Croydon, CR0 4YY

21/10/2024

01777088-0011